Forests and People

Forests and People

Property, Governance, and Human Rights

Edited by

Thomas Sikor and Johannes Stahl

First published in paperback 2024

First published 2011
by Earthscan

For a full list of publications please contact:
Earthscan
4 Park Square, Milton Park, Abingdon, Oxon OX14 4RN

and by Earthscan
605 Third Avenue, New York, NY 10158

Earthscan is an imprint of the Taylor & Francis Group, an informa business

Publisher's Note
The publisher has gone to great lengths to ensure the quality of this reprint but points out that some imperfections in the original copies may be apparent.

British Library Cataloguing in Publication Data
A catalogue record for this book is available from the British Library

Library of Congress Cataloging in Publication Data
Sikor, Thomas.
Forests and people : property, governance, and human rights / Thomas Sikor and Johannes Stahl.
p. cm.
Includes bibliographical references and index.
ISBN 978-1-84971-280-4 (hardback)
1. Forest people--Land tenure. 2. Forest people--Civil rights. 3. Forest people--Government relations. 4. Forest policy. 5. Forest conservation. 6. Forest management. I. Stahl, Johannes. II. Title.
GN394.S57 2011
306.09152--dc22
2011009518

ISBN: 978-1-84971-280-4 (hbk)
ISBN: 978-1-03-292413-7 (pbk)
ISBN: 978-0-203-12400-0 (ebk)

DOI: 10.4324/9780203124000

Typeset in Sabon
by Mapset, Gateshead, UK

Contents

List of Figures and Tables

Figures

Tables

List of Contributors

Shaunna Barnhart is an instructor at Penn State University, Pennsylvania, US, and is completing her PhD in geography. Her dissertation research, *"Just Like Nail and Flesh": Community Forestry, Biogas, and Environmental Governmentality in Nepal* was funded, in part, by a Fulbright Hays Fellowship, National Science Foundation Doctoral Dissertation Research Improvement Grant, and the Society of Woman Geographers. Her research explores the impact of biogas in rural Nepal, the role of community forestry in promoting biogas, and the implications of biogas in carbon trading markets.

Grazia Borrini-Feyerabend is coordinator of the ICCA Consortium (www.iccaforum.org), president of the Paul K. Feyerabend Foundation and vice chair for Europe of the Commission on Environmental, Economic, and Social Policy (CEESP) of the International Union for the Conservation of Nature (IUCN). Past head of IUCN Social Policy Program, she developed and chaired the IUCN professional networks that spearheaded policy advances on equity, governance of natural resources, and human rights in conservation. Among her over 20 volumes in several languages are *Bio-Cultural Diversity Conserved by Indigenous Peoples and Local Communities*; *Sharing Power*; *Indigenous and Local Communities and Protected Areas*; *Collaborative Management of Protected Areas; Beyond Fences*; and *The Wealth of Communities.*

Jessica Campese is an independent consultant, most recently focusing on the intersections of human rights and conservation. She was the IUCN CEESP focal point for human rights from 2006 to 2008, and managing editor of IUCN CEESP *Policy Matters 15: Conservation and Human Rights*. She also co-edited the book *Rights Based Approaches to Conservation: Exploring Issues and Opportunities for Conservation*, with co-editors Terry Sunderland, Thomas Greiber, and Gonzalo Oviedo.

Peter Cronkleton is an anthropologist with the Forest and Governance Program of the Center for International Forestry Research (CIFOR) and leader of CIFOR's global Smallholder and Community Forestry research program. His research has focused on forest tenure reform, rural social movements, and institutional governance in forest communities. He has recently published in the journals *Human Ecology*, *Forest Ecology and Management*, and *Society and Natural Resources*.

Stefan Dorondel is a researcher at the Francisc I Rainer Institute of Anthropology Bucharest and the Institute for Southeastern European Studies Bucharest. His current research is focused on changes in the post-socialist state and in the Romanian agrarian landscape. His book publications include *Water and Death: Funerary Rituals, Water Symbolism and the Other World Imaginary among Romanian Peasants*. He edited with Stelu Serban *Between East and West: Studies in Anthropology and Social History*.

Victoria M. Edwards is a partner of Taylor-Edwards Environmental Communications, a resource management research consultancy based in the UK. She has over 20 years of experience in research, consultancy, and university lecturing. She is the author of *Dealing in Diversity: America's Market for Nature*; *Corporate Property Management*; and numerous research papers and reports. She was awarded an OBE by Queen Elizabeth for services to the environment in 2004. She researches on the New Forest, southern England, and was involved in its designation as a national park.

Anne M. Larson is a senior associate with CIFOR and is based in Nicaragua. Her research has focused on local forest management, decentralization, indi-genous rights, land and forest tenure, and forest governance. Her book publications include the co-edited volumes *Forests for People: Community Rights and Forest Tenure Reform* and *Democratic Decentralization through a Natural Resource Lens*. She has authored several books in Spanish.

Godwin Limberg is project manager of the Fauna Flora International Murung Raya Conservation and Sustainable Development project in Central Kalimantan, Indonesia. Prior to this appointment, he collaborated as a consultant in CIFOR research on decentralization, community empowerment, poverty alleviation, and corporate social responsibility (CSR). An important focus of his work is community participation in resource management and balancing conservation and regional development. Limberg was co-editor of *The Decentralization of Forest Governance Politics: Economics and the Fight for Control of Forests in Indonesian Borneo*.

Fergus MacKay is senior counsel at the UK-based non-governmental organization (NGO) the Forest Peoples Programme. He has litigated a number of cases before the Inter-American Court of Human Rights and United Nations treaty bodies, including the 2007 *Saramaka People v. Suriname* case. His recent publications include "Indigenous peoples and international financial institutions", in D. Bradlow and D. Hunter (eds) *International Financial Institutions and International Law*; "From 'sacred commitment' to justiciable norms: Indigenous peoples' rights and the inter-American human rights system", in M. Salomon,

A. Tostensen, and W. Vandenhole (eds) *Casting the Net Wider: Human Rights and Development in the 21st Century*; and "Indigenous peoples' rights and the United Nations Committee on the Elimination of Racial Discrimination", in S. Dersso (ed) *Perspectives on the Rights of Minorities and Indigenous Peoples in Africa*.

Beth Rose Middleton is an assistant professor at the University of California, Davis, US, in the Department of Native American Studies, with a focus on Native environmental policy, geography, and economic/community development. Her book on Native land trusts, *Trust in the Land: New Directions in Tribal Conservation*, was published by University of Arizona Press in 2011 as part of the First Peoples series on New Directions in Indigenous Studies. She has also published articles in *Economic Development Quarterly*, *Journal of Political Ecology*, *Ethnohistory*, and *News from Native California*.

Moira Moeliono is a senior associate with the Forest and Governance Program of CIFOR. She has a long experience with community involvement in natural resource management, with a special interest in forest tenure and conservation issues. She has contributed to works on decentralization, community-based conservation, and tenure issues. More recently, she has been involved in research on national–local policy linkages in forest governance in the context of reducing emissions from deforestation and degradation (REDD).

Terry Parnell is the grassroots networking and advocacy advisor for the East West Management Institute's Program on Rights and Justice in Cambodia. With more than 15 years of experience in Cambodia, she works with a wide range of grassroots groups, including farmers, fishers, forest communities, and indigenous minorities as they seek to assert and protect their land and natural resource rights. Parnell has made recent presentations on Cambodian development and rights-related issues at the Woodrow Wilson Center in Washington, DC, and at Harvard University, US.

To Xuan Phuc is the Southeast Asia analyst for Forest Trends, based in Hanoi, Vietnam. He received his doctoral degree in geography at Humboldt University in Berlin in 2007, where he also worked as a junior researcher at the Junior Research Group on Post-Socialist Land Relations. His dissertation examines the political economy of the forest sector in Vietnam, with particular attention given to the dynamics of access and control over forestland and forest resources. From 2007 to 2009, he was a post-doctoral fellow at the Anthropology Department at the University of Toronto, Canada, where he was involved in a research project entitled Challenges of Agrarian Transitions in Southeast Asia (ChATSEA). Currently, in addition to his work with Forest Trends, he continues to work with ChATSEA and serves as a research fellow at the Asia Institute at the University of Toronto.

Blake D. Ratner is program leader, governance, at the WorldFish Center, and visiting senior research fellow at the International Food Policy Research Institute. An environmental sociologist, he conducts comparative research across natural resource systems, exploring the role of collective action in conflict prevention. His articles on accountability, equity, and ethics in environmental decision-making have appeared in the journals *Society and Natural Resources*, *Human Rights Dialogue*, *Population Research and Policy Review*, *Human Organization*, *Sociological Inquiry*, *Marine Policy*, and *Global Change, Peace and Security*.

Jesse C. Ribot is an associate professor of geography and director of the Social Dimensions of Environmental Policy Initiative at the University of Illinois (UI), US. Prior to joining UI, Ribot was a senior associate in the Institutions and Governance program at the World Resources Institute. He has been a fellow at the Max Planck Institute for Social Anthropology, a Woodrow Wilson International Center for Scholars fellow, a MacArthur fellow at the Harvard Center for Population and Development Studies, a fellow at the Yale Program in Agrarian Studies, lecturer in the Department of Urban Studies and Planning at the Massachusetts Institute of Technology (MIT), and has worked for numerous development agencies. He conducts research on decentralization and democratic local government; natural resource tenure and access; distribution along natural resource commodity chains; and vulnerability in the face of climate and environmental change.

Thomas Sikor has conducted research on forest tenure and politics over the past two decades, including fieldwork in Vietnam, Albania, and Romania. He holds a PhD in energy and resources from the University of California at Berkeley, US, and currently is reader in international development at the University of East Anglia, UK. Sikor has guest-edited a number of special journal issues and books on topics ranging from community and land reform to property theory. In his activist work, he promotes the expansion of economic, political, and cultural rights to disadvantaged people in Vietnam.

Neera M. Singh is an assistant professor at the Faculty of Forestry, University of Toronto, Canada. Her research focuses on democratization of forest governance, forest tenure and local rights, REDD regimes and equity, and the role of affect in forest–people relationships. Prior to her academic career, she founded and led a non-profit organization in India, *Vasundhara*, that works on community forestry and sustainable livelihood issues.

Johannes Stahl works on forest governance and biodiversity conservation. He was a Ciriacy-Wantrup post-doctoral fellow in Natural Resource Economics and Political Economy at the University of California at Berkeley, US, and holds PhD and MA degrees in agriculture and anthropology. His recent book, *Rent from the Land: A Political Ecology of Postsocialist Rural Transformation*, was published in 2010. He has also written articles for *Conservation and Society*, *Global Environmental Change*, and *Development and Change*. Stahl lives and works in Montreal, Canada.

William D. Sunderlin is principal scientist in the Forests and Livelihoods Program at CIFOR in Bogor, Indonesia. He is leader of research on REDD+ project sites in CIFOR's Global Comparative Study on REDD+ (reducing emissions from deforestation and degradation and enhancing carbon stocks). During 2006 to 2009, while working at the Rights and Resources Initiative in Washington, DC, he was lead author of a global survey of forest tenure entitled *From Exclusion to Ownership?: Challenges and Opportunities in Advancing Forest Tenure Reform*. The current themes of his research are rights, livelihoods, climate change, and the drivers of deforestation and forest degradation.

Peter Leigh Taylor is an associate professor of sociology and co-leader of the Environmental Governance Working Group at Colorado State University, US. He conducts research and applied work with community-based forestry and conservation organizations in Mexico and Central America. His recent articles have appeared in the *Journal of Rural Studies*, *World Development*, *Society and Natural Resources*, and *Sustainable Development*.

Preface

Over the past decade, not only have many governments heeded calls for forest tenure distribution by enacting legislative reforms, but advocacy for indigenous peoples' rights has also achieved groundbreaking successes on transnational agreements and court rulings, and international organizations have begun to develop new human rights-based approaches to forestry and conservation. Moreover, forest rights activists and philanthropic organizations have teamed up to strengthen global advocacy on forest people's behalf.

We present this book in the hope that it identifies key lessons and illuminates critical challenges for the unprecedented development of a global forest rights agenda. The book demonstrates, we believe, that forest rights activists share a common agenda and have advanced this agenda forcefully at the local, national, and global levels. At the same time, this agenda remains highly diverse, as it originates from different origins and retains an outstanding capacity to respond to specific contexts. It is this very responsiveness, facilitated by forest rights activists' capacities of critical reflection and mutual recognition, that has made the forest rights agenda so vibrant and relevant across heterogeneous contexts.

Given this global orientation, we intend this book, naturally, to also speak to issues related to REDD+ (reducing emissions from deforestation and degradation and enhancing carbon stocks). REDD+ has burst onto the scene over the past couple of years, threatening to usurp the entire forest agenda. Yet, as the chapters in this volume show, the relationship between forests and people goes beyond REDD+, and forest rights activists think beyond forest-based climate change mitigation.

This book originates from a workshop on forest rights held at the University of California at Berkeley, US, in May 2009. We called for the workshop to provide engaged scholars and critical activists an opportunity to reflect together on the emergent rights agenda in international forestry. Berkeley stands for an exceptional tradition of engaged scholarship on the relations between forests and people. Berkeley scholars have long been at the fore of research and advice on how to make forestry more just, whether in California, India, Indonesia, Zimbabwe, or other parts of the world. We felt extremely privileged to hold the workshop in Berkeley and have three outstanding exemplars of this tradition participate: Louise Fortmann, Nancy Peluso, and Jeff Romm.

Many people have contributed to this book, many more than we can acknowledge here. Our particular gratitude goes to our contributors, all the workshop participants, and to Don Reisman, our publisher at Earthscan, for their support, enthusiasm, and patience. Carol Colfer, Louise Fortmann, David Kaimowitz, Augusta Molnar, Jeff Romm, and Lini Wollenberg provided very helpful advice at various points during the production process. The German Research Council provided funding for the 2009 workshop under its Emmy Noether-Programm. Some of Johannes's writing time was covered under an S. V. Ciriacy-Wantrup Postdoctoral Fellowship at the University of California, Berkeley.

Thomas Sikor and Johannes Stahl
March 2011

List of Acronyms and Abbreviations

ACHR	American Convention on Human Rights
ACOFOP	Association of Forest Communities of the Guatemalan Petén
AfCom	African Commission on Human and Peoples' Rights
AIPRC	American Indian Policy Review Commission
CBD	Convention on Biological Diversity
CEESP	Commission on Environmental, Economic, and Social Policy
CERD	Committee on the Elimination of Racial Discrimination
CFUG	community forest user group
ChATSEA	Challenges of Agrarian Transitions in Southeast Asia
CI	Conservation International
CIBA	California Indian Basketweavers' Association
CIDOB	*Confederación de Pueblos Indígenas de Bolivia* (Confederation of Indigenous People of Eastern Bolivia)
CIFFMC	California Indian Forest and Fire Management Council
CIFOR	Center for International Forestry Research
CIHR	Conservation Initiative on Human Rights
CIRAD	French International Research Center for Agricultural Development
CLEC	Community Legal Education Center
CNS	*Conselho Nacional dos Seringueiros* (National Rubber Tappers' Council)
COPNAG	*La Central de Organizaciones de Pueblos Nativos Guarayos* (Central Organization of Native Guarayos Peoples)
CSR	corporate social responsibility
CWC	Central Women's Committee
3Es+	effectiveness, efficiency, equity, and co-benefits
EU	European Union
EWMI	East–West Management Institute
FAO	Food and Agriculture Organization of the United Nations
FECOFUN	Federation of Community Forest Users, Nepal
FFI	Fauna & Flora International
FLEG	Forest Law Enforcement and Governance process

FPIC	free, prior, and informed consent
FSC	Forest Stewardship Council
GIE	groupement d'intérêt économique (economic interest group)
IACHR	Inter-American Commission on Human Rights
ICCA	Indigenous Peoples' Conserved Territories and Areas Conserved by Indigenous Peoples and Local Communities
ICSO	Indigenous Community Support Organization
IRAM	Indigenous Rights Active Members network
IRD	French Institute for Development Research
IUCN	International Union for Conservation of Nature
JFM	joint forest management
KEF	Kalahan Educational Foundation
KNP	Kutai National Park
LLC	Local Land Commission (Romania)
MBR	Maya Biosphere Reserve
MIT	Massachusetts Institute of Technology
MLA	member of the legislative assembly
MMJSP	Maa Maninag Jungle Surakhya Parishad
MoF	Ministry of Forestry
MOU	memorandum of understanding
NAHC	California Native American Heritage Commission
NDF	National Department of Forestry (Romania)
NFNPA	New Forest National Park Authority
NGO	non-governmental organization
NTFP	non-timber forest product
OJM	Orissa Jungle Manch
PAE	agro-extractive settlement
PCNP	Piatra Craiului National Park
PCR	rural council president
PES	payment for environmental services
RAAN	North Atlantic Autonomous Region
RBA	rights-based approach
RECOFTC	Regional Community Forestry Training Center (recently renamed RECOFTC–The Center for People and Forests)
REDD	reducing/reduced emissions from deforestation and degradation
REDD+	reducing emissions from deforestation and forest degradation and the role of conservation, sustainable management of forests, and enhancement of forest carbon stocks in developing countries
RESEX	extractive reserve
RLC	Regional Land Commission (Romania)
TCO	*tierra comunitaria de origen* (original community land)
TNC	The Nature Conservancy
UI	University of Illinois

UK	United Kingdom
UNDRIP	United Nations Declaration on the Rights of Indigenous Peoples
UNFCCC	United Nations Framework Convention on Climate Change
US	United States
USFS	US Forest Service
WCS	Wildlife Conservation Society
WI	Wetlands International
WWF	World Wide Fund for Nature (*also known as* World Wildlife Fund)

1

Introduction: The Rights-Based Agenda in International Forestry

Thomas Sikor and Johannes Stahl

Rights have become a central concept in international forest policy and advocacy. Local people, forest communities, and indigenous peoples have long demanded tenure rights to forest, asserted cultural rights, and requested a say in their own affairs. Yet, it is only now that their individual demands as forest *people* and collective claims as forest *peoples* are finding recognition at national and global levels, with many governments enacting legislation to recognize customary tenure and governance; post-socialist governments in Europe and Asia restoring forests to their historical owners and distributing them to rural communities; the Inter-American Court of Human Rights ruling on cases dealing with forest peoples' collective rights to cultural and political self-determination; and negotiators for the United Nations Framework Convention on Climate Change (UNFCCC) making reference to the rights of indigenous peoples and the members of forest communities.[1]

The increasing prominence of this issue at national and global levels attests to the emergence of a rights agenda in international forestry. Demands for rights include three elements. Rights activists call for equity in the distribution of forest benefits, often in the form of redistribution of forest tenure. They advocate the recognition of forest people's particular identities, experiences, and visions. They also promote the participation of forest people in political decision-making regarding their own affairs. The rights agenda thereby demonstrates a strong orientation towards the goal of social justice, similar to the environmental justice movement in the US and global organizing around land

and food rights (cf. Schlosberg, 2004). It involves issues of property, governance, and human rights, as highlighted in the title of this book.

The attention to rights in forestry differs from efforts to implement "rights-based approaches" in international development, water management, and biodiversity conservation (cf. Scanlon et al, 2004; Campese et al, 2009; Hickey and Mitlin, 2009). There are three crucial differences. First, redistribution is a central demand of activists in forestry, but not in the other fields. Many forest rights activists are calling for not only the redirection of benefit streams to forest people, but also the redistribution of forest tenure to them upon the background of entrenched historical inequalities. Second, the rights agenda in forestry emerges from longstanding demands expressed by forest people in numerous grassroots initiatives. These grassroots foundations set it apart from rights-based approaches in development and conservation, which typically derive their legitimacy from transnational human rights norms and are largely driven by international and national organizations. Third, forest rights activists attend to people's individual rights as well as peoples' collective rights. In contrast, rights-based approaches in the other fields tend to emphasize individual rights.

This book offers a novel look at the emerging rights agenda in international forestry, and seeks to answer three questions. What is the rights agenda in international forestry? What are the key conceptual and strategic issues encountered by rights activists? What lessons have been learned on how to promote forest people's rights? To answer these questions the book brings together strategic analyses written by leading thinkers on international forest governance with a series of cutting-edge case studies contributed by young scholars. The chapters originate from Africa, Asia, Europe, Latin America, and North America, covering a wide variety of rights initiatives at local, national, and global levels.

In this introduction we begin with a brief description of the key demands made by forest rights activists, detecting unity in the diversity of concrete demands, actors, and actions constituting the rights agenda. This diversity, we argue, is due to the multiple origins of the rights agenda in organizing around tenure and indigenous rights, as well as recent human rights advocacy. Unity comes from a vision anchored in the goal of social justice shared by forest rights activists. Nevertheless, activists engage in vivid debates around four key issues, which we briefly synthesize here and which the contributions to the book explore in great detail: the kinds of claims to support; the sorts of actors considered to make legitimate claims; the types of authorities understood to recognize rights; and the political strategies serving state recognition of rights. The shared agenda, we conclude, puts activists in a strong position to confront global threats to forest people's rights. Yet, the need for global action also presents rights activists with new challenges – in particular, the need to sustain their capacity for reflexive recognition.

Unity in Diversity

There is a discernible unity to the forest rights agenda because calls for forest people's rights center on the three elements of equity, participation, and recognition. First, rights activists demand equity in the distribution of forest benefits. This demand often takes the form of appeals for a redistribution of forest tenure to redress the historical exclusion of people from forests. For example, rights activists in India and many countries of Latin America and sub-Saharan Africa highlight the need to acknowledge forest people's customary tenure rights in national legislation. Eastern European and East Asian governments have transferred significant parts of previously state-owned forests to rural people as part of post-socialist property reforms. In addition, demands for redistribution also take the form of calls for equitable sharing of benefits from logging, payments for environmental services, community–company partnerships, and actions on reducing emissions from deforestation and forest degradation and the role of conservation, sustainable management of forests, and enhancement of forest carbon stocks in developing countries (REDD+).

Second, rights activists call for recognition of forest people's identities, experiences, and visions. This is often expressed in terms of group identities, as forest peoples see themselves as outside the cultural mainstream and find their group-specific identities devalued. Activists such as the Indigenous Peoples Network of Malaysia point to the loss of diverse cultures due to a growing monoculture and to the need for recognition of social and cultural differences to overcome the stigmas attached to forest peoples. Calls for recognition also highlight forest people's individual experiences of social status, as they often find themselves positioned at the lower end of economic, social, and cultural relations. They demand the recognition of forest people as individuals and as full partners in social interactions – for example, motivating efforts to overcome entrenched gender differences. Moreover, recognition demands respect for forest people's visions of desirable lifestyles, economies, and forest landscapes. This aspect finds illustration in the attention given to traditional knowledge, as exemplified by efforts in Southeast Asia to validate shifting cultivation as a sustainable practice of land management.

Third, rights activists promote forest people's participation in political decision-making in matters that affect their own lives. They demand forest policy-making procedures that encourage public participation, democratic control over forests, and the conduct of local affairs in ways that involve community participation. They criticize forest agencies for excluding forest people from decisions about forests and for their lack of accountability to them. These efforts take many forms at the local, national, and global scale. They include demands for decentralizing forest management to elected local governments throughout the world and for the recognition of customary authorities in many countries, particularly in Africa and Latin America. They comprise efforts by rights activists in India and Nepal to promote forest people's participation in forest management. At the global level, they take the form of efforts to establish consultative forums for forest people's organizations regarding REDD+.

Rights activists often combine calls for equity, recognition, and participation. For example, the Forest Peoples Program connects advocacy for the recognition of forest peoples' distinct identities with demands to create particular avenues through which they can participate in political decision-making and to redistribute tenure. Many Indian activists tie actions aimed at recognition of forest people's low social status with efforts to get their demands for tenure redistribution acknowledged. Other rights activists focus their efforts on just one of the three above elements; for example, the Rights and Resources Initiative, a global coalition of advocacy, research, and philanthropic organizations, focused on the redistribution of forest tenure in its initial years.

Thus we find that there is a certain unity to the rights agenda in international forestry. Yet, this is unity in diversity, far from any uniformity. The concrete demands, actors, and actions constituting the rights agenda are tremendously diverse. This diversity finds reflection in this book, as the chapters provide abundant evidence for the variety of concrete demands made by forest people, peoples, and their supporters. The initiatives covered in the chapters involve many kinds of actors in support of forest people's claims, including international organizations, transnational networks, forest people's own associations, national governments, forest departments, social movements, professional non-governmental organizations (NGOs), and membership organizations. They also include various kinds of action in the pursuit of rights recognition, including direct action, everyday forms of resistance, advocacy, technical assistance, capacity-building, appeals to courts, regulatory reforms, legislative acts, law enforcement efforts, and administrative decisions.

Multiple Origins

The rights agenda in international forestry is so diverse in parts because the focus on forest people's rights stems from multiple origins. The agenda emerges from the confluence of three distinct sets of initiatives centered on calls for the redistribution of forest tenure, indigenous peoples' rights to self-determination, and human rights.

Forest activists have long advocated the equitable distribution of tenure rights to forests (e.g. Larson et al, 2010). They argue that the redistribution of forest tenure is necessary to redress people's historical dispossession from forests through nationalization and state management. The transfer of tenure to land and connected resources is the key strategy to overcome people's exclusion from forests. Tenure rights activists have included numerous civil society organizations, and their concerns have mirrored the central role attributed to tenure in academic research on forest management (e.g. Peluso, 1992). In their actions, activists have always maintained a strong connection with forest people's actions on the ground and the demands articulated by grassroots organizations.

The focus on local-level action finds reflection in the focus on providing practical support for forest people's claims on the ground, as illustrated by numerous "counter-mapping" initiatives. The call for transfer of tenure has recently gained significant momentum in many parts of the world, particularly in post-socialist countries and Latin America. How these successes may amount to a "tenure transition" is the subject of Chapter 2, contributed by William D. Sunderlin.

Indigenous peoples' organizations and their supporters have been demanding rights to political and cultural self-determination over the past three decades (e.g. Mander and Tauli-Corpuz, 2006). Their demands often include calls for the restitution of forest tenure to indigenous peoples, many of whom have used and managed forests historically. They also insist on indigenous peoples' participation in political decisions over their own affairs, including forest management. In contrast with tenure activists' strong focus on the grassroots, the proponents of indigenous peoples' rights have long-established strong associations, networks, and organizations at national and international levels. International NGOs such as the International Work Group for Indigenous Affairs, Almaciga, Rights and Democracy, and Tebtebba have played significant roles. Higher-level organizing has helped indigenous peoples and their supporters to lobby transnational bodies on indigenous rights, in particular the Permanent Forum on Indigenous Issues and the Inter-American Commission on Human Rights of the United Nations. It has also allowed them to successfully promote transnational agreements on indigenous rights, such as the 2007 United Nations Declaration on the Rights of Indigenous Peoples, and use transnational courts for their defense – for instance, bringing cases before the Inter-American Court of Human Rights. The prominence of legal advocacy in the indigenous people's movement is further discussed in Chapter 3 by Fergus MacKay.

The assertion of human rights is a more recent phenomenon in international forestry. Human rights activists and sympathetic conservationists seek to safeguard procedural and substantive rights in conservation. Procedural rights refer to a minimum level of participation in political decision-making, such as the right to information and access to justice. Substantive rights are about minimum standards of life that are considered commensurate with human dignity, including rights to life, health, food, housing, and work. References to human rights originate mostly from international organizations such as the International Union for Conservation of Nature (IUCN), and have relied on transnational bodies such as the now defunct United Nations Sub-Commission on the Promotion and Protection of Human Rights and the United Nations Human Rights Council. They focus on the formulation of "universal standards" and the development of global conventions, including legally binding human rights treaties (e.g. the International Covenant on Economic, Social, and Cultural Rights) and non-binding agreements (e.g. the United Nations Draft Declaration of Principles on Human Rights and the Environment). How human rights apply to forest conservation is discussed in further detail by Jessica Campese and Grazia Borrini-Feyerabend in Chapter 4.

All three sets of initiatives contribute to the emerging rights agenda in international forestry. Today's forest rights activists often have a background in organizing around tenure, indigenous peoples' demands, or human rights. Their demands reflect the claims made and experiences gained in all three fields. In consequence, the rights agenda in international forestry cannot be reduced to any of the three sets of initiatives. It goes beyond the call for tenure transfer, which has been the central rallying point for forest rights activists over many years, and extends to a larger set of forest people than those targeted by the supporters of indigenous peoples' rights. The rights agenda is also more encompassing than human rights-based approaches to conservation as it attends to the diversity of claims asserted by forest people and peoples at all levels.

A Unifying Vision of Social Justice

Forest rights activists have developed a unifying vision: activists of all kinds possess a strong commitment to social justice. Social justice provides a vision that helps them to bring together highly diverse demands, actors, and actions; to integrate multiple historical origins; and to make productive use of continuing debates among them. This vision prompts activists to demand the redistribution of forest tenure, forest people's participation in political decision-making, and the recognition of their identities, experiences, and aspirations.

Social justice provides the rights agenda with a powerful normative core that helps to keep centrifugal forces in check. Rights claims deserve recognition if they work towards "parity" as understood in an economic, social, cultural, and political sense (cf. Fraser, 2001). Forest people's claims find support if they help to overcome or reduce the entrenched inequalities characterizing forestry in many parts of the world. The principle of parity imposes requirements on the process through which claims are asserted and recognized as rights. Rights must result from democratic deliberative processes that are not skewed in favor of the interests of a dominant group. The principle also serves to distribute benefits, chances for recognition, and opportunities for political participation more equally. Claims and actors deserve support if recognition of their claims helps to ameliorate or erase some of the stark inequalities in the distribution of forest benefits. Similarly, forest rights activists appeal to particular authorities and employ certain strategies if these serve the recognition of claims asserted by the marginalized.

The focus on rights provides a unifying concept around which many activists and policy-makers can rally. Forest rights activists and policy-makers have successfully used the concept of "rights" for action at local, national, and global scales. Global activists call for the recognition of forest people's rights in the sense of a generalized notion of moral entitlement. Activists and policy-makers at the national level demand or legislate on the recognition of rights understood as legal relationships and procedures applicable in a uniform manner. Forest people and their supporters at the local level request concrete bundles of rights to forest resources and functions reflecting particular local histories and specificities. "Rights" thus serve as a unifying concept that facilitates not only coalitions between activists at any particular scale, but also connections across scales. Due to the plasticity of the term, rights function as the glue that keeps together a highly diverse set of demands, actors, and actions – despite the inherent tensions and open debate among activists.

The focus on rights makes the emerging agenda distinct from other established paradigms seeking to promote social justice in international forestry. At the risk of overgeneralizing, one can identify three other paradigms centered on the notions of stewardship, interests, and needs. The rights agenda differs from these, despite some overlaps. Calls for rights recognition are different from management approaches such as community forest management, co-management, and adaptive collaborative management, which argue for forest people's inclusion on the basis of their forest stewardship. The focus on rights also sets the rights agenda apart from efforts to increase forest people's participation in forest management, such as participatory and multi-stakeholder approaches, on the basis of their interests in forests. Calling for rights is also different from justifications for forest people's inclusion with reference to their needs under the overarching goal of alleviating poverty.

We even suggest that the focus on rights helps to avoid some of the conceptual and empirical problems confronted by the other paradigms. The rights agenda does not require the empirical assumption that local people are better forest stewards than other actors, which is difficult to uphold in practice (cf. Agrawal and Gibson, 1999). It puts the spotlight on the dramatic economic, social, cultural, and political inequalities characterizing international forestry, which often receive short shrift in approaches centered on the notion of multiple interests in forests (cf. Edmunds and Wollenberg, 2001). Neither does the focus on rights rest on problematic assumptions about the role of forests in poverty alleviation to justify forest people's inclusion on the grounds of their needs (cf. Sunderlin et al, 2005).

Furthermore, the chapters of this book demonstrate how the centrality of rights provides activists and policy-makers with guidance on concrete policies and strategies. The emphasis on rights and efforts to overcome existing inequalities allows activists to judge the potential of particular policies and strategies to serve the overarching goal of social justice. It helps them to distinguish devolution policies redistributing rights to disadvantaged forest people from those serving to exclude marginalized actors. It allows them to differentiate whether decentralization programs strengthen or undermine forest people's control over forests, and whether or not they enhance forest peoples' rights to self-determination. It also points out the difference between policies recognizing customary authorities that promote forest people's rights to political participation and those that reinforce the position of unaccountable leaders. The principle of parity also facilitates distinction between empowering and regressive forms of collaborative management.

Key Conceptual and Strategic Issues

Yet, forest rights activists also encounter challenging conceptual and strategic issues, issues that have been at the core of vibrant debates among activists about suitable ways to pursue rights recognition. In this book we highlight four key conceptual and strategic issues that underlie their debates. These issues have centered on four questions. What claims find support? Whose claims are considered to constitute rights? What authorities recognize forest people's rights? And what political strategies serve rights recognition by the state?

What claims find support?

There is no singular and uniform "right to forest". Instead, forest people assert a large repertoire of rights claims (e.g. Fortmann and Bruce, 1988). They demand rights as individual people, as various kinds of social groups and collectivities, as indigenous peoples, and as forest peoples. Their claims refer to specific forest resources, certain uses of forests, various kinds of environmental functions provided by forests, economic and cultural values associated with forests, or forests as a whole understood as bundles of resources and functions. They focus on legal tenure or extend this to the tangible and intangible benefits derived from tenure. Some emphasize demands for a minimum level of rights, whereas others call for egalitarian distribution or even request privileges for certain actors. This multiplicity of concrete claims poses a vexing challenge to activists and policy-makers, as there are many ways to translate the universal appeal to forest people's moral entitlements into concrete rights at global, national, and local levels. This multiplicity is the subject of Part II of this book, including contributions by Jesse C. Ribot and Anne M. Larson, Shaunna Barnhart, and To Xuan Phuc.

Whose claims are considered to constitute rights?

In many situations, multiple social actors assert their rights to forests (e.g. Colfer, 2005). Their claims may clash directly over certain forest resources, or they may be in conflict with one another as they relate to overlapping forest resources or functions (e.g. productive land versus protected areas versus indigenous territory). The claims may conflict at the local level as different kinds of social actors assert competing claims. New conflicts may arise over time as emerging new actors demand rights, or existing conflicts may lose relevance as some actors drop their claims. In addition, people and organizations located far from the forest may voice claims to specific forest resources or functions in competition with local actors. As a result, efforts to promote redistribution, recognition, or participation often require that activists and policy-makers make difficult choices about whose claims to consider to constitute rights, and whose justifications to support. These choices are discussed in Part III in the chapters by Moira Moeliono and Godwin Limberg, Victoria M. Edwards, and Beth Rose Middleton.

What authorities recognize forest people's rights?

Claims only become rights if they are sanctioned by authority (Sikor and Lund, 2009). Forest people reference their claims to a variety of institutions and procedures considered to possess authority, and often to more than one. Conversely, in most situations, plural authorities endorse claims to forests in overlapping or even competing ways. The institutions include democratically elected local governments, customary arrangements, central governments, courts, transnational conventions, and social norms. This pluralism of institutional authority challenges activists and policy-makers to identify the institutions and procedures that are most conducive to the recognition of forest people's rights. This is the focus of Part IV in the book, with contributions by Anne M. Larson and Peter Cronkleton, and Stefan Dorondel.

What political strategies serve rights recognition by the state?

Forest people and their supporters employ a variety of political strategies to promote redistribution, recognition, and participation. Their efforts involve tactical and strategic choices about the concrete rights asserted, the forums used to voice claims, the organizations and associations established to promote forest people's rights, and the coalitions to be formed in the pursuit of rights recognition. This wide repertoire of available strategies opens up many opportunities for activists and decision-makers to pursue rights recognition, yet it also forces them to identify the most promising strategies in specific political contexts. These strategies are the topic of Part V, including chapters by Neera M. Singh, Blake D. Ratner and Terry Parnell, and Peter Cronkleton and Peter Leigh Taylor. Global strategies are also covered in Chapters 2 to 4.

The debates led by forest rights activists around these questions attest to the vibrancy and relevance of the rights agenda in international forestry. The issues have provided the grounds for productive tensions among activists, becoming an important source of dynamism over recent years. These tensions have allowed the rights agenda not only to sustain its rich diversity, but also to maintain sufficient flexibility for context-specific engagement. They have only been productive, however, because activists have shared a capacity for reflexive recognition. Forest rights activists have not sought to resolve all the key issues once and for all. Instead, they have acknowledged differences among themselves and used these to sustain deliberative processes about forest rights and the strategies best employed to pursue rights recognition.

This capacity for reflexive recognition has opened up productive entry points for engaged scholarship. Research on forest rights, whether academic, policy-oriented, or applied, has made significant contributions to the work of forest rights activists. Forest rights activists' capacity for reflection also provides a key rationale for the present book. At the broadest level, we have conceived the book to contribute to critical and constructive reflection among rights activists. More specifically, it is intended to generate important insights for key strategic choices faced by activists. Which claims deserve activists' support? Which actors should they assist? To which authorities should they appeal in their pursuit of rights recognition? Which strategies should they employ?

Challenges of Globalization

Rights activists increasingly face the need to respond to global threats to forest people's rights arising from changes in global commodity prices and regulatory regimes. The demand for cheap energy drives the conversion of large tracts of forestland to biofuel plantations. Payment for environmental services (PES) schemes attribute new market values to forests and make them available for individual appropriation. The inclusion of REDD+ in post-2012 climate architecture translates the carbon storage capacity of forests into monetary values. These market and regulatory changes transform the monetary values attributed to forests and assign new ones, raising the attractiveness of forests to powerful global actors as a potential source of accumulation.

Forest rights activists are in a strong position to counter global forces and to influence the appropriation of increased forest values in forest people's favor. They are well equipped to critically engage with ongoing market and regulatory changes at a conceptual and practical level. Conceptually, the rights agenda questions a preoccupation with efficiency in policy-making if this becomes divorced from concerns over distribution. It also calls into question the lack of attention to entrenched inequalities and political contestations over forests in market-based thought. At a more practical level, the rights agenda provides activists and policy-makers with strategic guidance on the potential of market-based policies to lead to socially just forestry. It allows them to critically engage the promotion of community-based enterprises, the commercialization of forest products, and integration within international commodity markets, as indicated in several chapters. In particular, the rights agenda positions activists in a strong position to encounter REDD+, as we seek to show in the concluding chapter.

At the same time, the urgency of developing global strategies challenges rights activists in many ways. At a conceptual level, they find themselves confronted with efforts to appropriate their agenda by actors who do not share the goal of redistribution, as has occurred in community-based natural resource management (Brosius et al, 2005). Attempts to appropriate the rights agenda in part originate from calls for forest tenure reforms advanced by international organizations (e.g. World Bank, 2003). While sharing the concern over forest tenure, these calls promote a narrow set of rights – typically, individual exclusive ownership – and deflect attention away from the inequalities entrenched in forest tenure. Attempts at appropriation also come from human rights approaches that focus on procedural rights, such as application of the principle of free, prior, and informed consent, or limiting substantive rights to a minimum level of livelihood benefits (e.g. Wilson, 2009). Rights activists will have to find ways to counter such attempts at appropriation by emphasizing the overarching goal of social justice and stressing notions of rights and authority that are often incompatible with the interests of the powerful.

At a more strategic level, forest activists are increasingly facing global forces that operate at scales reaching far beyond the established forums for forest advocacy and policy-making. The focus on forest rights has lent activists significant strength due to a clearly focused agenda, in particular through the emphasis on the redistribution of a particular asset (forests). Recent experience with global negotiations over the role of forests in climate change mitigation, however, indicates the limits of such a specialized agenda. Forest rights activists watched from the sidelines as climate change negotiators turned to forests and put REDD+ back on the table, and they continue to be overwhelmed by the apparent expedience of forest-based climate change mitigation. Such experiences indicate the political advantages of a broader rights agenda that develops strategic alliances with related movements. Forest rights activists will face new debates about the value of potential alliances, particularly with broader agrarian movements and advocacy on climate justice.

Finally, the urgency of global advocacy challenges the core strength of the rights agenda in international forestry: the combination of unity with diversity. Efforts to counter global threats confront decisions about who gets to establish definitions of forest people's rights and represent these in global forums. Global activists see an increasing need to create processes of representation in order to develop capacities for global responses. At the same time, they recognize the inherent problems of attempts to represent a highly diverse set of concrete demands, actors, and actions at local and national levels. Global activists face the difficult task of representing forest people in global forums without ignoring these problems inherent in their representation. Activists at all levels face the challenge of instituting mechanisms that keep global activists accountable to forest people and activists at local and national levels without inhibiting or fragmenting global efforts.

The need to globalize thus challenges the capacity for reflexive recognition that has nurtured the vibrancy and relevance of the rights agenda in international forestry to date. Only by maintaining this capacity will activists be able to sustain unity in diversity. It is also here where engaged scholarship can make critical contributions to the forest rights agenda: illuminate key issues encountered by activists in their attempts to develop a global agenda. This book, therefore, presents scholarly analyses that engage practical concerns and take a constructive stance. It stands for scholarship that respects forest people's and activists' claims, enters into a dialogue with them, facilitates critical reflection on past experiences, and points out future possibilities. The underlying premise is that such scholarship is critical to the pursuit of socially just forestry.

Acknowledgements

We thank Thomas Enters, Louise Fortmann, Fergus MacKay, Jeff Romm, and Peter Leigh Taylor for inspiring and constructive comments. Sally Sutton provided competent editorial assistance. This chapter has also benefited from discussions at the workshop Towards a Rights-Based Agenda in International Forestry?, held at the University of California at Berkeley; the Global Environmental Justice Seminar at the University of East Anglia; the workshop Taking Stock of Smallholder and Community Forestry: Where Do We Go from Here?, organized by the Center for International Forestry Research (CIFOR), the French Institute for Development Research (IRD), and the French International Research Center for Agricultural Development (CIRAD); and the STEPS Conference 2010: Pathways to Sustainability at the Institute of Development Studies.

Note

1 Hereafter we use the term "forest people" to refer to the many kinds of people and peoples making claims on forests, for reasons of convenience. Our use of the term does not want to deny its unfavorable connotations in particular contexts or to ignore the actual diversity of actors, as the following discussion will make clear.

References

Agrawal, A. and C. C. Gibson (1999) "Enchantment and disenchantment: The role of community in natural resource conservation", *World Development*, vol 27, no 4, pp629–649

Brosius, J. P., A. Lowenhaupt Tsing, and C. Zerner (eds) (2005) *Communities and Conservation: Histories and Politics of Community-Based Natural Resource Management*, AltaMira Press, Walnut Creek, CA

Campese, J., T. Sunderland, T. Greiber, and G. Oviedo (eds) (2009) *Rights-Based Approaches: Exploring Issues and Opportunities for Conservation*, CIFOR and IUCN, Bogor

Colfer, C. J. P. (ed) (2005) *The Equitable Forest: Diversity, Community & Resource Management*, Resources for the Future, Washington, DC

Edmunds, D. and E. Wollenberg (2001) "A strategic approach to multistakeholder negotiations", *Development and Change*, vol 32, no 2, pp231–253

Fortmann, L. and J. Bruce (eds) (1988) *Whose Trees? Proprietary Dimensions of Forestry*, Westview Press, Boulder, CO

Fraser, N. (2001) "Recognition without ethics?", *Theory, Culture & Society*, vol 18, no 2–3, pp21–42

Hickey, S. and D. Mitlin (eds) (2009) *Rights-Based Approaches to Development: Exploring the Potential and Pitfalls*, Kumarian Press, Sterling, VA

Larson, A. M., D. Barry, G. Ram Dahal, and C. J. P. Colfer (eds) (2010) *Forests for People: Community Rights and Forest Tenure Reform*, Earthscan, London

Mander, J. and V. Tauli-Corpuz (eds) (2006) *Paradigm Wars: Indigenous Peoples' Resistance to Globalization*, Sierra Club, Washington, DC

Peluso, N. (1992) *Rich Forests, Poor People: Resource Control and Resistance in Java*, University of California Press, Berkeley, CA

Scanlon, J., A. Cassar, and N. Nemes (2004) *Water as a Human Right?*, IUCN, Gland, Switzerland

Schlosberg, D. (2004) "Reconceiving environmental justice: Global movements and political theories", *Environmental Politics*, vol 13, no 3, pp517–540

Sikor, T. and C. Lund (2009) "Access and property: A question of power and authority", *Development and Change*, vol 40, no 1, pp1–22

Sunderlin, W., A. Angelsen, B. Belcher, P. Burgers, R. Nasi, L. Santoso, and S. Wunder (2005) "Livelihoods, forests, and conservation in developing countries: An overview", *World Development*, vol 33, no 9, pp1383–1402

Wilson, E. (2009) *Company-Led Approaches to Conflict Resolution in the Forest Sector*, The Forests Dialogue, New Haven, CT

World Bank (2003) *Land Policies for Growth and Poverty Reduction*, World Bank, Washington, DC

Part I

Global Perspectives

The forest rights agenda originates from the confluence of three distinct sets of initiatives, as we have noted in Chapter 1. The initiatives have centered on calls for the redistribution of forest tenure, indigenous peoples' rights of self-determination, and human rights, respectively. Forest rights activists' visions and demands continue to reflect these distinct origins.

The chapters in Part I set out global perspectives on forests rights from the viewpoint of these distinct origins. The perspectives are global in the sense that they provide an overview of rights-related developments worldwide. In addition, their global perspectives derive from reference to institutions at the global level, such as global conventions, transnational law, and international courts, and to global-scale processes – for example, the liberalization of capital markets and increasing offshore sourcing of agricultural and biofuel crops. This global orientation sets the chapters apart from the remaining ones in the book, as the latter tend to emphasize practices, strategies, and processes at the local or national level.

Sunderlin reviews recent changes in forest tenure in Chapter 2. There is a "tenure transition" under way, he suggests, characterized by a discernible shift away from comprehensive government control of forests towards greater access and ownership by forest people. The key underlying causes of the tenure transition are the failure of state forest management, decentralization, resource management devolution, the decrease of natural forest timber rents in some countries, democratization, and international solidarity campaigns. Nevertheless, the prospects of further tenure reform are unclear, as new demands for forestland – such as those for offshore agricultural production, biofuels, and forest carbon – constitute a potentially grave threat to continued transition.

MacKay charts the judicial recognition of indigenous peoples' rights to land in transnational law in Chapter 3. Emphasizing the intimate relationship between land tenure regimes and indigenous peoples' demands for collective self-determination, he expounds upon the right of indigenous peoples to give or withhold free, prior, and informed consent as related in the Inter-American Human Rights System and the Inter-American Court of Human Rights. Landmark court rulings and key transnational agreements on indigenous rights indicate that land tenure regimes are currently undergoing considerable re-evaluation – offering increasing support to indigenous people's collective tenure to traditional territories. Specific court rulings acknowledge and articulate indigenous peoples' rights, as they assign considerable decision-making and governance powers to traditional territory holders, including the right to control territories through indigenous laws, customs, traditions, and institutions.

Campese and Borrini-Feyerabend discuss human rights-based approaches to conservation in Chapter 4. Indigenous peoples' and local communities' organizations and non-governmental organizations (NGOs) have increasingly embraced approaches to conservation that seek to safeguard basic human rights for affected people. These human rights-based approaches hold the potential to integrate conservationists' agendas and right-holders' priorities, as illustrated by initiatives developed within the International Union for Conservation of Nature (IUCN) and its Commission on Environmental, Economic, and Social Policy (CEESP). Yet, Campese and Borrini-Feyerabend also point out that practitioners need to acknowledge and deal with pre-existing power relations if new approaches are to be more than rhetorical tools. They stress that effective human rights-based approaches imply fairer and more sustainable governance systems that address inequities through time. Practitioners wishing to implement their conservation agendas in ways more respectful of right-holders can take advantage of international policy advances on governance, such as new guidance by the IUCN and the Convention on Biological Diversity (CBD) regarding Indigenous Peoples' Conserved Territories and Areas Conserved by Indigenous Peoples and Local Communities (ICCAs).

Taken together, the three chapters demonstrate forest rights activists' shared commitment to social justice, as we have highlighted in Chapter 1. Whatever their intellectual and personal origins, forest rights activists all thrive to overcome the entrenched economic, political, and cultural inequalities characterizing forested areas in many parts of the world. Social justice provides a unifying vision that provides the required kit for overcoming apparent differences in emphasis.

The chapters also indicate clear overlaps and connections between the three sets of initiatives. As forest rights activists have been able to make productive use of inherent tensions, they have learned from each other and come to acknowledge the benefits of cooperation and exchange. For example, Sunderlin mentions international human rights campaigns as a factor facilitating the tenure transition. MacKay highlights land tenure as a key concern to mobilizations for indigenous peoples' rights. Campese and Borrini-Feyerabend note the synergy between indigenous peoples' mobilizations and the efforts of sympathetic conservationists to develop human rights-based approaches to conservation.

Nevertheless, the three chapters demonstrate how the different origins continue to influence the forest rights agenda. The agenda remains highly diverse, as activists operationalize the shared commitment to social justice in different ways. It is far from uniform, as we noted in Chapter 1, as it has achieved unity in diversity. The distinct origins show up not only in the global perspectives but also in many of the local and national-level analyses presented in the remainder of the book. For example, Barnhart, Larson and Cronkleton, and Dorondel examine local and national instances of tenure transitions in Chapters 6, 11, and 12. Middleton, and Cronkleton and Taylor speak to issues of particular pertinence to indigenous peoples in Chapters 10 and 15. Moeliono and Limberg report attempts to develop a human rights-based approach to forest conservation in Indonesia in Chapter 8.

2

The Global Forest Tenure Transition: Background, Substance, and Prospects

William D. Sunderlin

In the course of recent decades, there has been a tendency for some governments to relax control over their national forest estate and to increase the area used or owned by non-state entities, such as communities, indigenous peoples, individuals, and firms. This broad pattern, called the forest tenure transition, is happening at a slow pace and is quite uneven among countries, yet it is measurable (White and Martin, 2002; Sunderlin et al, 2008).

This chapter aims to answer four questions:

1. What is the global forest tenure transition?
2. What forces explain the widespread assertion of government control over forests and the suppression of the rights of forest peoples dating back centuries, and what explains the partial restoration of those rights in recent decades?
3. Is there a basis for supposing that the tenure transition is meaningful to the interests of forest peoples given the slow and uneven pace of change?
4. What is the prospect that the forest tenure transition will continue?

These questions are answered in a necessarily cursory manner given space limitations. Reference is made to resources that expand upon these answers.

The subsequent sections address each of these four questions in turn. The chapter closes with a summary and conclusion.

What Is the Global Forest Tenure Transition?

A recent study shows that during the period of 2002 to 2008, in 25 of the 30 most forested countries accounting for 80 percent of the global forest estate, the absolute area of forestland under the control of governments decreased (from 80.3 to 74.3 percent of the forest area), and there was a corresponding increase in the area of land designated for use by communities and indigenous groups (from 1.5 to 2.3 percent of the area), land owned by communities and indigenous groups (from 7.7 to 9.1 percent of the area), and land owned by individuals and firms (from 10.5 to 14.2 percent of the area) (see Figure 2.1) (Sunderlin et al, 2008, pp7–10). A follow-up study found that in the same period, in 30 of 39 tropical forest countries accounting for 47 percent of the global forest estate and 85 percent of the tropical forest area, roughly the same pattern of tenure change is evident (RRI and ITTO, 2009, pp10–13). Among continental regions there is considerable unevenness in the transition, with 36 percent of the area of public forestlands administered by governments in Latin America, 68 percent in Asia, and 98 percent in Africa (RRI and ITTO, 2009, pp15–16). These studies build on a research framework introduced by White and Martin (2002).

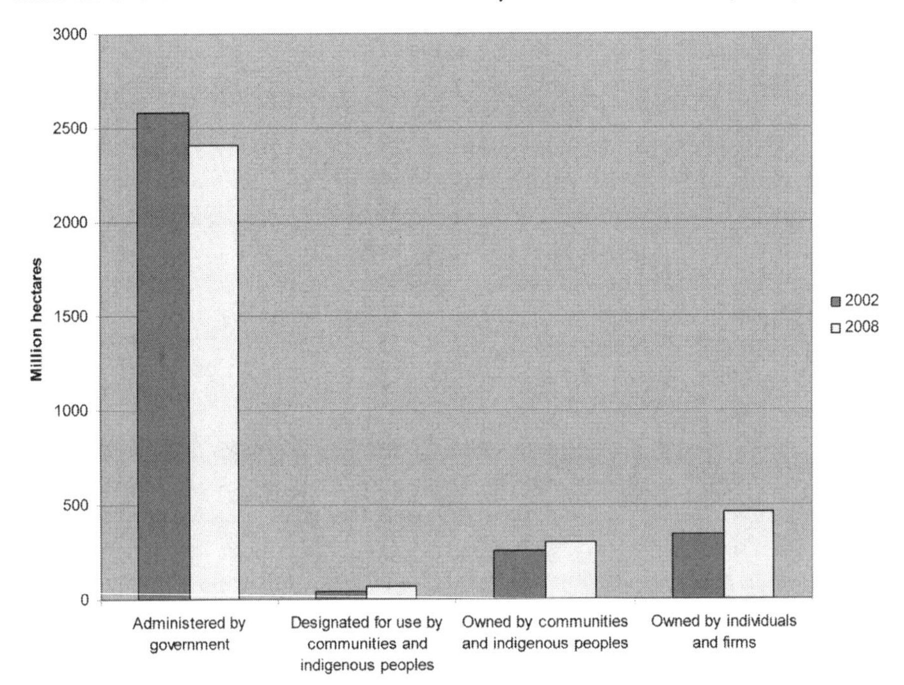

Figure 2.1 *Forest tenure distribution by tenure category in 25 of the 30 most forested countries, 2002–2008*

Source: Sunderlin et al (2008, p9)

What Forces Explain the Suppression and Restoration of Rights?

Customary and statutory rights

The forest tenure transition is essentially a movement from relatively robust local control over forest lands and resources by local people in an early historical stage, followed by suppression of the rights of forest peoples as governments assert control over national forests, and then (in some national cases) a relaxation of government control over a portion of the national forest estate and formal recognition of some local claims over forest lands and resources. What are the forces that explain these changes? In order to answer that question, we need to first unpack the term "rights".

Forest tenure is the right, whether defined in customary or statutory terms, that determines who can hold and use forest lands and resources, for how long, and under what conditions. Customary rights are generally determined at the local level by forest dwellers themselves. Through the lens of customary tenure, forest peoples are the rightful owners and managers of all forests because they have ancestral claims that predate the arrival of the state. Statutory (sometimes called "formal" or "legal") rights are determined by the state and were created, in part, to assert government control over and against customary claims. These two systems of rights have been in mutual opposition and contestation for centuries. The forest tenure transition represents, in many cases, a partial yielding of control by governments and at least partial recognition of customary claims by local people.

It is important to bear in mind that the data on the forest tenure transition measure only statutory rights, and not customary rights over forestland. These data are therefore far from being the last word on forest rights because forest peoples continue to assert their claim to a far larger share of the global forest estate than has yet been ceded.

Reasons for suppression of rights

In some ways, the suppression of the rights of forest people is a simple story. Originally, forest dwellers lived largely outside the reach of the nation state. Elaboration of their own tenure and management systems had relatively free rein for lack of interference by government authority and commercial exploitation. In the course of recent centuries, however, with the advent of capitalism, the unfolding of the industrial revolution, the establishment of cities and centers of trade, the growth of economic production and consumption, and the marketing of primary goods both domestically and internationally, emerging nation states and commercial entities asserted themselves forcefully and decisively in the forest biome. Kingdoms and colonial powers made formal decrees appropriating forests in the name of the greater good. States were not hesitant to resort to violence and intimidation in cases where there was resistance to seizing forest lands and resources, as in the case of India (Guha, 1989; Guha, 2001, p216) and German East Africa (Sunseri, 2003). Through the practice of indirect rule, colonial powers created a cadre of domestic administrators for the imposition of their will, as in the case of Uganda (Turyahabwe and Banana, 2008, p643) and Zimbabwe (Mandondo, 2000, p5). Many laws and treaties were deployed to establish state hegemony in the forest domain and to restrict the rights of forest peoples (Lynch et al, 1995, pp33–45). From the time of the seizure of forests by governments, protest movements by local people have ensued in Europe and colonized countries (Guha, 2001, pp213, 217).

Various trends deepened these broad economic, political, and institutional transformations, increased demand on forest lands and resources, and underlay the suppression of rights:

- rapid population growth and the emergence of a consumer class, both within and outside a given country (i.e. increase in the number of consumers and the amount consumed per person) leading *inter alia* to expanded conversion of forestlands to agriculture and the exploitation of timber and other forest resources;
- imperialism and colonialism, with rich nation states establishing access to foreign lands and resources in poorer nation states for the production of commodities specific to the tropics and unavailable in the home country;
- systems of patronage whereby the state conferred privileged access to timber and other resources to itself, elites, members of the commercial class, and the military;
- the growth of infrastructure and roads to support commercial activities linking cities and ports to the hinterland;
- coercive relations of production facilitating the control and expansion of resource-based commercial activities;
- competition among imperial powers and commercial rivalries in colonial trade, motivating the equation of forest control with defense of national interests, and motivating the creation of a war-making apparatus to defend national territory and resources (e.g. the need for timber to create a naval fleet);
- the imposition of systems of statutory tenure and a cadastral system, in part to lay the basis for fiscal forestry (forest taxation) (Scott, 1998, pp11–48);
- the role of not just economic, but also political, cultural, and religious imperialism in justifying the suppression of the rights of forest peoples (colonial laws and regulations prohibiting customary forest management practices asserted that indigenous rural peoples – including forest peoples – are backward, uneducated, destructive of natural resources, and in need of guidance towards enlightened modern beliefs, behavior, and practices).

There are cases where, centuries ago, representatives of national governments justified the imposition of government control and the suppression of local rights in the belief that local people destroy forests, and that therefore the law must intercede to protect forests.[1] In many national contexts, this reasoning proves to be weak and insubstantial when viewed in the context of the underlying factors listed above, and when taking into account that forest destruction by governments and commercial entities is often greater than that by local peoples.[2] There is sometimes an element of truth in government claims that local forest practices can be destructive, especially in cases where rapid local population growth finds no outlet in opportunities for rural to urban migration. Yet, in many cases, the accusation of destructive behavior by forest peoples serves as a pretext for deflecting attention away from more fundamental reasons for the disappearance of forests that put the state in a poor light.

Reasons for restoration of rights

Against this backdrop, and in view of the fact that there continue to be strong international and national interests favoring privileged government and commercial control over the forest landscape, it is fitting to ask: why have some governments relaxed control over some forests, recognized some customary land claims, and partly restored rights lost long ago? There are five main reasons: the failure of government control of the forest estate; decentralization and resource management devolution; the removal of timber rents from forests; democratization; and international human rights campaigns.

Failure of government control. Beginning in the 1970s and 1980s, national governments in forested countries, particularly in the developing world, took note of the fact that heavy-handed government control of forest lands and resources was a failure. Government forest management plans promised sustainable timber production, and protection of watersheds, biodiversity, and natural areas; yet, in many cases governments were presiding over rapid deforestation and forest degradation, depletion of timber stocks and related decrease in foreign exchange revenue, massive soil erosion and damage to agricultural production, siltation of rivers, and flora and fauna species extinctions. Recognizing these problems, national policy-makers were increasingly open to new approaches that involved decentralization of the forest-sector authority to the local level, and devolution of forest management to the community.

Decentralization and devolution. The impetus for forest-sector decentralization and devolution was, in part, the failure of central government control of the forest sector as described above; but it is only part of the explanation. In some cases, a shift towards decentralization and devolution long precedes the collapse of government efforts to manage forests. Also, the trend towards decentralization in developing countries is propelled by various factors outside the forest sector that can induce forest-sector decentralization. During the 1950s and 1960s, decentralization was carried out in developing countries to facilitate the transition from colonial status to independence, and was a means of improving planning and implementation of national development, especially rural development (Conyers, 1983, pp98–99). During the 1980s decentralization is mainly motivated by the aim to increase economic efficiency in the allocation of goods and services, and is associated with the neoliberal reform agenda and structural adjustment programs as a response to the failures of the centralized state (Crawford and Hartmann, 2008, p12). During the 1990s, decentralization is associated with the "good governance" reform agenda, democratization as a means of popular empowerment, poverty reduction, and conflict management (Crawford and Hartmann, 2008, pp12–13). There is a consensus between left and right on the desirability of decentralization, with the political left aiming for community empowerment at the local level, and neoliberals seeking fiscal constraint and shrinking of the centralized state (Crawford and Hartman, 2008, p12).

Within the forest sector itself, the most common goals of decentralization in practice have been to reduce costs, increase forest department revenues, affirm private property rights, and address legitimacy issues related to economic and political crises of central government (Larson, 2005, p34). It should be noted that forest-sector decentralization and devolution sometimes fail to produce improvements for the betterment of the lives of forest peoples (Edmunds and Wollenberg, 2001, pp190, 194; Edmunds and Wollenberg, 2003; Christy et al, 2007, p83; Duncan, 2007; Sikor and Nguyen, 2007) or for forest management (Tacconi, 2007). Tellingly, forest-sector decentralization can lead to righting of historical wrongs; but in no case is that the main reason why central governments embarked on decentralization (Larson, 2005, p34).

Removal of timber rents. In some countries the main historical impetus behind the assertion of government control over forestlands is the mandate to ensure privileged access to natural timber stands for commercial exploitation and foreign exchange revenue. It is arguable, though it is not yet empirically proven, that the disappearance of natural timber supplies underlies diminished state interest in central government control of forests. There is circumstantial evidence to support this notion. In some (but not all) cases, degraded forests are those that are assigned by national governments through devolution for community forestry (Edmunds and Wollenberg, 2001, p192). In other words, the removal of the high value content of forests makes it feasible for governments to recognize local claims to the forest. It should be noted, however, that some forests bereft of natural timber are still quite valuable for alternative uses (conversion to agriculture or silviculture) or for resources other than timber (e.g. mining).

Democratization. In recent decades, many developing countries have experienced a shift from authoritarian to democratic rule. This often opens up political space for asserting claims over lands and resources and for expressing dissent over past injustices and violations of rights. The avenues for assertion of property claims are strongest in countries where a range of rights are defended in law and in practice (e.g. free speech, right to assembly, civil and human rights) and where there is a functioning judicial system or other means for the public redress of grievances.

International human rights campaigns. International human rights campaigns and solidarity movements for cultural survival and sovereignty have proven important to the partial restoration of rights. Northern solidarity with indigenous peoples and ethnic minorities has had a role in improving human rights via external pressure on state policy.

Regional particularities

As noted above, about one third of the whole forest estate in Latin America is administered by government, about two-thirds in Asia, and almost 100 percent in Africa (RRI and ITTO, 2009, pp15–16). This suggests that there may be historical and geographic factors particular to a region which explain its status with respect to the transition. Why does Latin America appear to be far ahead of Asia and Africa with respect to the transition? Given the current state of knowledge, it is not possible to answer this question with certainty. Certainly, part of the answer is that Brazil occupies a large share of the regional forest estate, and conditions in that country have tended to favor a relatively strong civil society role in forest management. There are some interesting hypotheses as to why Brazil, in particular, and Latin America, in general, are ahead in the transition:[3]

- A wide area of national territory in Brazil was never fully colonized.
- The colonial presence ended longer ago in Latin America than in many African and Asian countries.
- There are strong institutions in support of democratization in Latin America and there is comparatively large political space for resistance.
- There were very low population densities in the Amazon Basin until 1970, with Amerindians as the only occupants. Relative to settlement patterns in Africa and Asia, Amerindians did not face much competition from other indigenous and peasant groups.[4]
- When road-building and colonization programs began in the Brazilian Amazon, the Amerindians (with missionary assistance) began a process of "ethnogenesis" which strengthened their claim to lands in the face of invasions.[5]
- The granting of extractive reserves and indigenous reserves in the Brazilian Amazon was the result of effective organizing with effective external pressure.[6]
- There has been a strong international solidarity movement in support of various indigenous peoples in Latin America.
- The timber sector has been weak in Latin America compared to Asia.

Is the Transition Meaningful?

The negative view

An argument can be made that the forest tenure transition has been insubstantial. The narrative would go something like this. No matter how much progress has been made during the period of 2002 to 2008, it remains the case that governments control three-quarters of the area of the global forest estate. The process of change is slow, and possibly slower than the numbers suggest.[7] While eighteen countries experienced decrease in the area of forest controlled by government in 2002 to 2008, only ten countries experienced increase in the area of land designated for use by communities and indigenous peoples, only seven countries experienced an increase in the area of forestland owned by communities or indigenous peoples, and only five countries experienced an increase in forestland owned by individuals or firms (Sunderlin et al, 2008, p9). Only eight countries account for almost all of the net increase in the area of lands designated for and owned by communities and indigenous peoples (Sunderlin et al, 2008, p10). The strong regional skew of the transition (much progress in Latin America, some in Asia, and almost none in Africa) raises doubts about the prospect of a transition in many countries.

It is not clear that conditions in many developing countries are conducive to betterment of the lives of forest peoples through the strengthening of rights to local lands and resources. In some country cases, formal designation of use rights or even ownership rights are not enforced and do not enable the right-holder to effectively exclude competing claimants. In some developing countries, weaknesses in governance (e.g. low capacity, lack of funds, corruption, lack of political will) undermine the ability to carry out forest tenure change, or to realize gains once tenure change has been achieved.

Perhaps the greatest source of concern is the additional threat to existing rights from new types of demand for access to lands and resources. This has been called the "global land grab" by rights advocates (RRI, 2010, pp15–17) and a "rise of global interest in farmland" by the World Bank (2010). Whereas there was an average annual global expansion of less than 4 million hectares prior to 2008, there were 45 million hectares of farmland deals announced before the end of 2009 (World Bank, 2010, pvi). The World Bank (2010, pvi) downplays the significance of this figure, saying much of this planned investment will not be realized; but this view is contested by rights advocates (GRAIN, 2010).

This trend is being propelled, in part, by offshore agricultural production in cases where there is no home-country agricultural frontier, the expansion of agro-industrial and silvicultural plantations, livestock ranching, and conservation efforts. Peak oil and incipient transformation of world energy supplies are inducing rapid expansion in the production of biofuels, as well as exploitation of fossil fuels in remote areas as more accessible supplies are exhausted. The prospect of marketing forest carbon in REDD (reducing emissions from deforestation and degradation) schemes for climate change mitigation has unknown consequences for forest rights. In most cases, rights to forest carbon have not yet been defined and are likely to be sought zealously by governments and corporate entities. Forest peoples are largely unaware of this new frontier for resource exploitation in their midst.

The positive view

A different argument can be made (supported by this author) that the forest tenure transition, in spite of its limitations and weaknesses, is an important breakthrough that ought to be defended and carried forward.

Without a doubt, the forest tenure transition reverses a historical pattern in some countries and is a moral victory worthy of celebration. Inasmuch as the suppression of the rights of forest people often involved brazen theft and brutal suppression, the increasing state recognition of local forest property rights aims in the direction of restitution, even though it might come nowhere near to fully compensating past injustices and losses. From the viewpoint of forest peoples themselves, recognition of customary rights lost long ago is an end in itself, not just a means to an end.

What concretely has rights restoration produced in the way of measurable and tangible benefits for the livelihoods and well-being of forest peoples? On the one hand, there is evidence of positive gains from resource management devolution and community forestry, which is predicated on strengthened local tenure. For example, case studies in Mexico suggest not just livelihood but environmental gains from community forest enterprises (Klooster, 2003; Bray, 2005; Bray et al, 2008). There is a widespread, emerging consensus in multilateral and bilateral donor organizations that strengthened tenure rights can confer a wide range of benefits, including poverty alleviation, reversal of environmental damage, and strengthening of local economic growth (FAO, 2002; Deininger, 2003; DFID, 2007; SIDA, 2007). Nevertheless, the alleged concrete benefits of strengthened local rights remain to be demonstrated more conclusively.

One of the most positive aspects of the forest tenure transition is that it is occurring at a time when claims on forest lands and resources may be escalating and changing form. If forest peoples are to navigate these new waters successfully, they need to benefit from a trend that clearly – even if only weakly so far – favors their long-term interests.

Is the Transition Durable?

What is the prospect that the forest tenure transition will continue? There is no reason to suppose that trend is destiny. There are factors that favor the continuation of the trend, and factors that undermine it. We can suppose that, ultimately, in each country, the story will be told as a net outcome of competing forces, some of them with deep roots in history, and others that are entirely new and an abrupt departure from the past.

The following are among the main factors that tend in the direction of weakening local forest rights in the future. Although some of the original catalytic forces have disappeared or greatly weakened (e.g. colonialism, religious imperialism, etc.), others not only persist but increase the threat to rights. Notable among these are increased external demand for forest lands and resources in most forested countries, together with the persistence of a coalition of urban-based interests (the state, military, elites, corporations) to get access to these resources, and the extension of infrastructure and road growth to ensure that access. Longstanding structural patterns and trends underlie these factors: population growth, the growth of the middle class and consumer culture, and both national and corporate competition for control of access to lands and the marketing of primary goods. Although the exploitation of natural timber does not figure as prominently in countries where marketable stands have been eliminated, it may grow in importance in countries where timber is still abundant. As stated earlier, new forms of demand for forestland (offshore food production, biofuels and other agro-industrial and silvicultural products, forest-based fossil fuels, forest carbon) together constitute a major new frontier and, therefore, a potentially grave threat. Partial restoration of rights was motivated, in part, by state acknowledgement of past management failure; but reassertion of state control is plausible in cases where new forms of forest investment are judged vital to the national interest.

Against this picture there are factors that favor persistence and possible increase of efforts to strengthen the rights of forest peoples. The trend towards decentralization and resource management devolution will persist in many countries, though as pointed out earlier, this does not always guarantee favorable outcomes for local peoples. The success of devolution will depend on the degree of collective action by forest peoples to realize their own goals (Agrawal and Ostrom, 2001, p101) and on policies addressing structural inequities (Larson et al, 2007a, 2007b). The momentum towards worldwide democratization will likely continue, though there are imaginable scenarios where democratic gains can be eroded, particularly in situations where economic conditions deteriorate and states rescind liberties in the name of national stability. It is likely that international human rights campaigns and solidarity with indigenous peoples will grow in scope and influence. The global economic recession has slowed some forms of land-based investments, probably averting some conflict over forestlands; but eventual restoration of global economic growth is likely to restore these investments to former levels. Diversification of some economies out of the rural sector, together with rural to urban migration, may reduce conflict on forestlands and brighten the rights picture. Under a conceivable arrangement of circumstances, the establishment of REDD projects could improve the situation of local rights. Social and environmental safeguards, including free, prior, and informed consent (FPIC) and prior resolution of tenure ambiguities, are a requirement for the establishment of REDD projects and for third-party certification of those projects. It is by no means a foregone conclusion, however, that tenure ambiguities and conflicts will be resolved, or resolved in favor of forest peoples.

Conclusions

During recent years there has been a discernible shift away from wholesale government control of forests towards greater access and ownership by people who live in forests. However, three-quarters of the area of the global forest estate remains under government control, and the forest tenure transition has proceeded very unevenly among continental regions and countries.

Hundreds of years ago, governments asserted control over world forests and restricted the rights of local people. Among the main motivating factors were growing demand for land and resources in relation to the development of capitalism, industrialism, and economic imperialism; urbanism; population increase; rivalries among nodes of colonial and mercantile power; and also political, cultural, and religious imperialism.

During recent decades the transition towards partial recognition of rights in some countries has been set in motion by the failure of government control as an approach to forest management, decentralization and resource management devolution, the decrease of natural forest timber rents in some countries, democratization, and international solidarity campaigns.

On balance, the global forest tenure transition is an important historical development for both moral and practical reasons. Although the transition has not yet fulfilled the rights aspirations of forest peoples on the scale that it could or should, it nevertheless represents an important achievement well worth pushing forward. Under conditions where old and new demands on forest resources loom large, the need to strengthen local rights is ever more necessary.

It is impossible to forecast whether the transition will continue. On the one hand, the forces that gave rise to the transition for the most part persist, and there are emerging pressures on governments and corporate entities to strengthen the tenure rights of forest peoples. On the other hand, some of the forces that originally led to the loss of rights continue, and there are entirely new forms of demand on forest resources, on an unprecedented scale, that conceivably could erase the gains that have been won.

Acknowledgements

I am very grateful to the following people who contributed ideas and information that have found their way into this chapter. They are Deborah Barry, Carol Colfer, Pam Jagger, David Kaimowitz, Anne Larson, John Lindsay, Esther Mwangi, Jesse Ribot, J. Timmons Roberts, Tom Rudel, Liz Alden Wily, and Lini Wollenberg. I alone am responsible for any errors of fact or interpretation in this chapter.

Notes

1 See, for example, the case of Dutch colonial forest policy in Indonesia being driven by the assumption that customary land tenure and use are inappropriate and destructive (Galudra and Sirait, 2009).
2 See, for example, the case of the Madras Presidency in India during the late 18th century (Saravanan, 2003).
3 I am indebted to the following people for generously offering me the ideas which helped produce this list: Carol Colfer, Anne Larson, John Lindsay, Jesse Ribot, J. Timmons Roberts, Tom Rudel, and Liz Alden Wily.
4 Pers comm, Tom Rudel, 2 March 2009.
5 Pers comm, Tom Rudel, 2 March 2009.
6 Pers comm, J. Timmons Roberts, 2 March 2009.
7 The 2002 figures are the latest data available for that year, and do not necessarily reflect on-the-ground conditions during that year. They may, in fact, reflect conditions from a prior year. As a result, the measured change in a given country might take place over a period longer than six years.

References

Agrawal, A. and E. Ostrom (2001) *Collective Action, Property Rights, and Devolution of Forest and Protected Area Management*, DSE/ZEL, Feldafing, Germany

Bray, D. B. (2005) "Community forestry in Mexico: Twenty lessons learned and four future pathways", in D. B. Bray, L. Merino-Pérez, and D. Barry (eds) *The Community Forests of Mexico: Managing for Sustainable Landscapes*, University of Texas Press, Austin, TX, Chapter 14, pp335–349

Bray, D. B., E. Duran, V. H. Ramos, J.-F. Mas, A. Velazquez, R. B. McNab, D. Barry, and J. Radachowsky (2008) "Tropical deforestation, community forests, and protected areas in the Maya Forest", *Ecology and Society*, vol 13, no 2, p56, www.ecologyandsociety.org/vol13/iss2/art56

Christy, L. C., C. E. Di Leva, J. M. Lindsay, and P. T. Takoukam (2007) "Decentralization and devolution", in L. C. Christy, C. E. Di Leva, J. M. Lindsay, and P. T. Takoukam (eds) *Forest Law and Sustainable Development*, World Bank, Washington, DC, Chapter 7, pp83–100

Conyers, D. (1983) "Decentralization: The latest fashion in development administration?", *Public Administration and Development*, vol 3, pp97–109

Crawford, G. and C. Hartmann (2008) "Introduction: Decentralisation as a pathway out of poverty and conflict?", in G. Crawford and C. Hartmann (eds) *Decentralisation in Africa: A Pathway out of Poverty and Conflict?*, Amsterdam University Press, Amsterdam, The Netherlands, Chapter 1, pp7–32

Deininger, K. (2003) *Land Policies for Growth and Poverty Reduction*, World Bank and Oxford University Press, Washington, DC

DFID (UK Department for International Development) (2007) *Land: Better Access and Secure Tenure for Poor People*, DFID, London

Duncan, C. R. (2007) "The impact of regional autonomy and decentralization on indigenous ethnic minorities in Indonesia", *Development and Change*, vol 38, no 4, pp711–733

Edmunds, D. and E. Wollenberg (2001) "Historical perspectives on forest policy change in Asia: An introduction", *Environmental History*, vol 6, no 2, pp190–212
Edmunds, D. and E. Wollenberg (eds) (2003) *Local Forest Management: The Impacts of Devolution Policies*, Earthscan, London and Sterling, VA
FAO (Food and Agriculture Organization of the United Nations) (2002) *Land Tenure and Rural Development*, FAO Land Tenure Studies 3, FAO, Rome, Italy
Galudra, G. and M. Sirait (2009) "A discourse on Dutch colonial forest policy and science in Indonesia at the beginning of the 20th century", *International Forestry Review*, vol 11, no 4, pp524–533
GRAIN (2010) *World Bank Report on Land Grabbing: Beyond the Smoke and Mirrors*, http://farmlandgrab.org/15542
Guha, R. (1989) *The Unquiet Woods: Ecological Change and Peasant Resistance in the Himalaya*, University of California Press, Berkeley, CA
Guha, R. (2001) "The prehistory of community forestry in India", *Environmental History*, April, pp213–238
Klooster, D. (2003) "Campesinos and Mexican forest policy during the twentieth century", *Latin American Research Review*, vol 38, no 2, pp94–126
Larson, A. M. (2005) "Democratic decentralization in the forestry sector: Lessons learned from Africa, Asia and Latin America", in C. J. P. Colfer and D. Capistrano (eds) *The Politics of Decentralization: Forests, Power and People*, Earthscan, London and Sterling, VA, Chapter 2, pp32–62
Larson, A. M., P. Pacheco, F. Toni, and M. Vallejo (2007a) "The effect of forestry decentralization on access to livelihood assets", *The Journal of Environment and Development*, vol 16, no 3, pp251–268
Larson, A. M., P. Pacheco, F. Toni, and M. Vallejo (2007b) "Trends in Latin American forestry decentralisations: Legal frameworks, municipal governments and forest dependent groups", *International Forestry Review*, vol 9, no 3, pp734–747
Lynch, O. J., K. Talbott, and M. S. Berdan (1995) *Balancing Acts: Community-Based Forest Management and National Law in Asia and the Pacific*, World Resources Institute, Washington, DC
Mandondo, A. (2000) *Situating Zimbabwe's Natural Resource Governance Systems in History*, CIFOR Occasional Paper no 32, Center for International Forestry Research, Bogor, Indonesia
RRI (Rights and Resources Initiative) (2010) *The End of the Hinterland: Forests, Conflict and Climate Change*, RRI, Washington, DC
RRI and ITTO (Rights and Resources Initiative and International Tropical Timber Organization) (2009) *Tropical Forest Tenure Assessment: Trends, Challenges and Opportunities*, RRI, Washington, DC, and ITTO, Yokohama, Japan
Saravanan, V. (2003) "Colonial commercial forest policy and tribal private forests in Madras Presidency: 1792–1881", *Indian Economic & Social History Review*, vol 40, no 4, pp403–423
Scott, J. C. (1998) *Seeing Like a State: How Certain Schemes to Improve the Human Condition Have Failed*, Yale University Press, New Haven, CT, and London
SIDA (Swedish International Development Cooperation Agency) (2007) *Natural Resource Tenure*, Position paper, SIDA, Stockholm, Sweden
Sikor, T. and T. Q. Nguyen (2007) "Why may forest devolution not benefit the poor: Forest entitlements in Vietnam's Central Highlands", *World Development*, vol 35, no 11, pp2010–2025

Sunderlin, W. D., J. Hatcher, and M. Liddle (2008) *From Exclusion to Ownership?: Challenges and Opportunities in Advancing Forest Tenure Reform*, Rights and Resources Initiative, Washington, DC

Sunseri, T. (2003) "Reinterpreting a colonial rebellion: Forestry and social control in German East Africa, 1874–1915", *Environmental History*, vol 8, no 3, pp430–451

Tacconi, L. (2007) "Decentralization, forests and livelihoods: Theory and narrative", *Global Environmental Change*, vol 17, pp338–348

Turyahabwe, N. and A. Y. Banana (2008) "An overview of history and development of forest policy and legislation in Uganda", *International Forestry Review*, vol 10, no 4, pp641–656

White, A. and A. Martin (2002) *Who Owns the World's Forests?: Forest Tenure and Public Forests in Transition*, Forest Trends and Center for International Environmental Law, Washington, DC

World Bank (2010) *Rising Global Interest in Farmland: Can It Yield Sustainable and Equitable Benefits?*, World Bank, Washington, DC

3

Indigenous Peoples' Rights and the Jurisprudence of the Inter-American Human Rights System

Fergus MacKay

How to ensure the effective protection of biological diversity and forests while also respecting indigenous peoples' rights is an issue currently facing conservation groups, governments, and indigenous peoples. This is particularly the case in connection with the establishment and management of protected areas and proposed reduced emissions from deforestation and degradation (REDD) projects. A large percentage of existing protected areas worldwide are on lands traditionally owned by indigenous peoples (Amend and Amend, 1992), and proposed REDD projects, which seek to maintain or increase carbon storage in standing forests, will disproportionately affect indigenous peoples (Dooley et al, 2008; Griffiths, 2008). The vast majority of these protected areas have been established and/or managed in violation of indigenous peoples' rights. The same may also be the case in connection with extant and planned REDD projects.

This chapter focuses on indigenous peoples' rights as elaborated in the Inter-American Human Rights System, especially by the Inter-American Court of Human Rights (the Court). As noted in Chapter 1 of this book, the focus is on collective rights or the rights of peoples, rather than individual rights. It begins by looking at rights to traditional territories and the right to give or withhold free, prior, and informed consent (FPIC). This is followed by a discussion of the Court's jurisprudence on restitution and a pending case that explicitly seeks restitution of indigenous lands from existing protected areas. In this sense, the demands made by indigenous peoples, which are increasingly recognized and affirmed in international jurisprudence, go beyond a call for a simple reassessment of benefit-sharing measures. As discussed in Chapter 1, these demands and the associated jurisprudence embrace a substantial and durable re-evaluation of tenure regimes in order to fully recognize and respect indigenous peoples' collective tenure – derived from their own laws, customs, and traditions – as well as their rights to control their traditional territories effectively through their own institutions. This includes the right to participate effectively in decision-making as it may affect their territories and the right to give or withhold their consent.

Rights to Lands, Territories, and Resources

The primary organs of the Inter-American Human Rights System are the Inter-American Court of Human Rights and the Inter-American Commission on Human Rights (IACHR). These bodies supervise compliance with two main human rights instruments: the 1948 American Declaration on the Rights and Duties of Man and the 1969 American Convention on Human Rights (ACHR). The IACHR, which issues recommendations rather than binding decisions, is competent to receive complaints about alleged violations of the American Declaration (applicable to all members of the Organization of American States, or OAS) and the ACHR (applicable to 25 of the 34 OAS member states). The Court is competent to adjudicate contentious cases provided that the respondent state is a party to the ACHR and has accepted its jurisdiction (22 states have accepted the Court's jurisdiction). The decisions of the Court are binding on respondent states and may be executed in domestic courts.

In their decisions and judgments, the Court and the IACHR have repeatedly held that indigenous peoples' property rights derive from their own laws, land tenure systems, and their traditional occupation and use, and that these rights are valid and enforceable irrespective of formal recognition by national law (IACtHR, 2001, para 164; IACtHR, 2006, para 248; IACHR, 2002, para 130–1; IACHR, 2004, para 117). In the 2006 *Sawhoyamaxa Indigenous Community* case, for instance, the Court observed that its jurisprudence holds that "traditional indigenous land ownership is equivalent to full title granted by the State", and includes the "right to demand official recognition of their property and its consequent registration" (IACtHR, 2006, para 128, p248). These norms have been reaffirmed and further elaborated upon by the Court in five judgments issued between 2005 and 2010 (IACtHR, 2005a, 2005b, 2006, 2007, 2010).

In the 2005 *Moiwana Village Case*, the Court recognized important norms in relation to displaced persons and communities – an issue that is highly relevant to protected areas, especially in Africa (Redford and Fearn, 2007). In particular, it held that many of the *1998 United Nations Guiding Principles on Internal Displacement* are illustrative of state obligations under the ACHR, and emphasized, accordingly, that "States are under a particular obligation to protect against the displacement of indigenous peoples, minorities, peasants, pastoralists and other groups with a special dependency on and attachment to their lands" (IACtHR, 2005b, para 111). This is broadly consistent with Article 10 of the 2007 United Nations Declaration on the Rights of Indigenous Peoples, which strictly prohibits forced relocation, and which has been used as an interpretive aid previously by the Court and others (IACtHR, 2007, para 131; CERD, 2008, para 29). The jurisprudence of the IACHR also holds that consent is required in cases involving relocation of indigenous peoples (IACHR, 1984, p120).

The norms summarized above are not unique to the Inter-American System. Indigenous peoples' rights to own and control their traditional territories are also protected in similar terms under United Nations human rights instruments. The Committee on the Elimination of Racial Discrimination (CERD), for example, has called on state parties to "recognize and protect the rights of indigenous peoples to own, develop, control and use their communal lands, territories and resources" (CERD, 1997, para 4). It routinely reaffirms this principle when examining state reports and in communications adopted under its urgent action and early warning procedures.

The Convention on Biological Diversity (CBD), an international environmental treaty currently in force for 193 states, also addresses indigenous peoples' rights, including in relation to the establishment and management of protected areas. Decision VII/28 on Protected Areas, adopted by the Conference of Parties to the CBD, provides that "the establishment, management and monitoring of protected areas should take place with the full and effective participation of, and full respect for the rights of, indigenous and local communities consistent with national law and applicable international obligations" (CBD, 2004, para 22). These applicable international obligations are defined, *inter alia*, in international human rights law, including in the jurisprudence of the IACHR and the Court.

Article 10(c) of the CBD further provides that state parties shall "protect and encourage [indigenous peoples'] customary use of biological resources in accordance with traditional cultural practices that are compatible with conservation or sustainable use requirements". This article, by implication, should also be read to include protection for rights to lands and resources and to require recognition and protection of indigenous institutions and customary laws relating to ownership, use, and management of biological resources. These conclusions are supported by the analysis of the Secretariat of the CBD and by indigenous peoples' own research on the measures needed to implement and give effect to Article 10(c) (CBD, 1997). Conducted in five countries, this research demonstrates that respect for indigenous peoples' institutions and customary laws is intrinsic to biodiversity and ecosystem protection and that secure land tenure rights and control over traditional territory are critical elements of effective conservation and use of biological diversity (Colchester, 2006).

The Right to Effective Participation and Free, Prior, and Informed Consent

In line with United Nations human rights bodies (CESCR, 2001, para 12; CERD, 2003, para 16; HRC, 2006) and mechanisms (e.g. SRRF, 2009, pp13–15), the IACHR has consistently held that indigenous peoples' free, prior, and informed consent is required in relation to activities that may affect their traditionally owned territories (IACHR, 2002, para 131; IACHR, 2004, para 142; IACHR, 2006a, para 214). The IACHR explained in this respect that consultation with indigenous peoples:

> *... must be designed to secure the free and informed consent of these peoples ...; must guarantee participation by indigenous peoples ... in all decisions on natural resource projects on their lands and territories, from design, through tendering and award, to execution and evaluation[; and] ... must also ensure that such procedures will establish the benefits that the affected indigenous peoples are to receive, and compensation for any environmental damages, in a manner consistent with their own development priorities.* (IACHR, 2007a, para 248)

The Court has also developed jurisprudence on FPIC. In 2007, in *Saramaka People v. Suriname*, the Court held that indigenous peoples hold the right to self-determination (IACtHR, 2007, para 93) and, consequently, have the right "to freely determine and enjoy their own social, cultural and economic development" (para 95). This, in turn, requires that recognition of their territorial rights must also include recognition of "their right to manage, distribute, and effectively control such territory, in accordance with their customary laws and traditional collective land tenure system" (para 194). These enumerated and extensive powers in relation to traditional territory have clear implications for protected areas and REDD projects. Yet, they are routinely ignored in protected area design and management decisions, often to the detriment of effective protection and management measures themselves (Sobrevila, 2008).

Having ruled that the Saramaka people have protected property rights to their traditionally owned territory and the traditionally used natural resources therein (para 122), the Court examines whether states may restrict those rights, pursuant to Article 21(1) of the American Convention (which provides that the "law may subordinate use and enjoyment [of property] to the interest of society") (para 122–40). On the basis of this provision, the Court has previously held that states may restrict the use and enjoyment of the right to property only "where the restrictions are: (a) previously established by law; (b) necessary; (c) proportional; and (d) with the aim of achieving a legitimate objective in a democratic society" (para 129).

While the same criteria apply to proposed restrictions to indigenous peoples' property rights, provided such restrictions may be classified as "exceptional"

(IACtHR, 2008, para 49) the Court holds that an additional "crucial factor to be considered is whether the restriction amounts to a denial of their traditions and customs in a way that endangers the very survival of the group and of its members" (IACtHR, 2007, para 128). Survival in this context is much more than physical survival, however, and, instead, concerns each peoples' ability to maintain their multiple relationships to traditional territory (IACtHR, 2008, pp36–37) so that "they may continue living their traditional way of life, and that their distinct cultural identity, social structure, economic system, customs, beliefs and traditions are respected, guaranteed and protected" (IACtHR, 2007, para 121; IACtHR, 2008, para 37). The Court explains that for the state to guarantee that restrictions do not endanger the Saramaka people's survival, the state must comply with the following:

- It must ensure "effective participation" with regard to any development or investment project within their territory. A "development or investment" is defined by the Court as "any proposed activity that may affect the integrity of the lands and natural resources within the territory of the Saramaka people" (and thus clearly includes protected areas and REDD projects) (IACtHR, 2007, para 129, note 127).
- The Saramaka people must receive a reasonable benefit from the project.
- Independent environmental and social impact assessments must be undertaken (IACtHR, 2007, para 129).
- The state must implement "adequate safeguards and mechanisms in order to ensure that these activities do not significantly affect traditional lands and natural resources" (IACtHR, 2007, para 158).

For some developments or investments, "effective participation" requires the state to not only consult with indigenous peoples, "but also to obtain their free, prior, and informed consent, according to their customs and tradition" (IACtHR, 2007, para 134, 137). These investments are characterized as "large-scale" projects that may have a major impact within indigenous peoples' territories (IACtHR, 2007, para 137). In its 2008 interpretation judgment, the Court modifies the "major impact" requirement and explains that FPIC is required for any large-scale project that "could affect the integrity" of indigenous peoples' lands and territories (IACtHR, 2007, para 17), including the cumulative impact of small projects (IACtHR, 2007, para 41).

The African Commission on Human and Peoples' Rights (AfCom) heavily relied on *Saramaka People* in its 2010 decision in the *Endorois Welfare Council v. Kenya*. This case concerned the displacement and exclusion of the Endorois from a protected area. In addition to demonstrating that indigenous peoples' rights are actionable in African human rights institutions and reinforcing the universal nature of those rights, AfCom holds that "In terms of consultation, the threshold is especially stringent in favour of indigenous peoples, as it also requires that *consent* be accorded" (AfCom, 2010, para 226). Echoing *Saramaka*, it adds that with respect to "any development or investment projects that would have a major impact within the Endorois territory, the State has a duty not only to consult with the community, but also to obtain their free, prior, and informed consent, according to their customs and traditions" (AfCom, 2010, para 291).

The Court and AfCom are the first human rights bodies to make a distinction between situations where FPIC is required and those in which a lower standard of participation is required. The Committee on the Elimination of Racial Discrimination, for instance, holds that "no decisions directly relating to [indigenous peoples'] rights and interests are taken without their informed consent" (CERD, 1997, para 4(d)). Likewise, the Human Rights Committee has "stress[ed] the obligation of the State party to seek the informed consent of indigenous peoples before adopting decisions affecting them" (HRC, 2006, para 22). However, in its case law under the Optional Protocol I, while the committee has held that "participation in the decision-making process must be effective, which requires not mere consultation but the free, prior and informed consent of the members of the community", it has done so only in relation to "the admissibility of measures which substantially compromise or interfere with the culturally significant economic activities of a minority or indigenous community" (HRC, 2009, para 7.6). While limited in scope, this has clear implications for the establishment and management of the majority of protected areas and proposed REDD measures, and, importantly, its geographical scope is universal rather than regional.

The Right to Restitution

In international law, violation of a human right gives rise to a right of reparation for the victim(s) (van Boven, 1993). Reparations include restitution, compensation, rehabilitation, satisfaction, and guarantees of non-repetition (van Boven, 1993, p57). In the 2005 *Yakye Axa Indigenous Community* case, the Court first addressed indigenous peoples' right to the restitution of traditional lands. It determined that a violation of the right to property had occurred because Paraguay had failed to effectively restore and secure the rights of the Yakye Axa to their traditional lands, large parts of which were held by private persons. It ordered that the state identify these traditional lands and establish a fund for the expropriation and return of privately held lands, free of charge, to the Yakye Axa (IACtHR, 2005a, para 217).

Similar violations were also found in the *Sawhoyamaxa Indigenous Community* case (IACtHR, 2006, para 248). In this case, the Court explains that indigenous peoples maintain their property rights in cases where they have been forced to leave or have otherwise lost possession of their lands, including where their lands have been expropriated or transferred to third parties, unless this was done in good faith and consensually (IACtHR, 2006). It also examined the temporal scope of indigenous peoples' right to restitution and held that it continues as long as indigenous peoples maintain some degree of connection with the lost lands (IACtHR, 2006, para 131). Evidence of the requisite connection may be found in "traditional spiritual or ceremonial use or presence; settlement or sporadic cultivation; seasonal or nomadic hunting, fishing or harvesting; use of natural resources in accordance with customary practices; or any other factor characteristic of the culture of the group" (IACtHR, 2006). The Court further held that if indigenous peoples are prevented by others from maintaining their relationships with their territories, the right to recovery continues "until such impediments disappear" (IACtHR, 2006, para 132).

If a state is unable to return indigenous peoples' traditional lands and communal resources for "concrete and justifiable reasons", compensation or the provision of alternative lands is required (IACtHR, 2006, para 138–139). In such cases, the Court requires that indigenous peoples' consent be obtained "in accordance with their own consultation processes, values, uses and customary law", with regard to choices about the provision of compensation or alternative lands (IACtHR, 2005a, para 151; IACtHR, 2006, para 135).

While neither the IACHR nor the Court has yet applied the above jurisprudence to a case involving protected areas, this is only a matter of time as more than one case is currently pending before the IACHR that directly requests restitution of indigenous lands incorporated within protected areas (one is discussed in detail below) (e.g. IACHR, 2006b). When resolving these cases, the Court and IACHR will likely be persuaded by the jurisprudence of the CERD, in addition to applying the above-cited norms enumerated by the Court. This is all the more likely given that the provisions utilized by CERD (protecting property and participation rights) employ similar language to that found in relevant provisions of the ACHR.

CERD has articulated two main interrelated rules applicable to the establishment of nature reserves in indigenous peoples' territories. First, in 2002, it held that "no decisions directly relating to the rights and interests of members of indigenous peoples be taken without their informed consent" in connection with a nature reserve in Botswana (CERD, 2002, para 304). Second, with regard to a national park in Sri Lanka, the committee called on the state to "recognize and protect the rights of indigenous peoples to own, develop, control and use their communal lands, territories and resources" (CERD, 2001, para 335). More generally, CERD has recognized that indigenous peoples have a right to restitution of their traditional territories and resources, which in principle also applies to nature reserves (CERD, 2006, para 17).

In 2007, CERD elaborated further, expressing its concern "about the consequences for indigenous groups of the establishment of national parks in the State party and their ability to pursue their traditional way of life in such parks" (CERD, 2007, para 22). It recommends that the state provide detailed information:

> *... on the effective participation of indigenous communities in the decisions directly relating to their rights and interests, including*

> *their informed consent in the establishment of national parks, and as to how the effective management of those parks is carried out.* (CERD, 2007, para 22)

The jurisprudence summarized above recognizes that indigenous peoples have extensive rights to own and control their traditional territories pursuant to the interrelated rights to self-determination, property, culture, participation, and others. This includes the right to restitution of lands taken without consent provided that indigenous peoples maintain some form of relationship therewith. Where lands cannot be returned for objective reasons, there is a right to compensation and the provision of other lands of equal quality and quantity (most indigenous peoples would assert that only their traditional lands, however, are of the same quality). While the state may, as an exceptional measure, restrict these rights, it must comply with a series of prior conditions, including securing indigenous peoples' effective participation in decision-making from the inception. This will also mean obtaining FPIC in most, if not almost all, cases. There is no reason to doubt that such rights would also apply in the context of the establishment and management of protected areas or REDD projects. Indeed, the Court has held that these rights apply to "any proposed activity that may affect the integrity of the lands and natural resources" within indigenous and tribal peoples' territories.

The Case of the Kaliña and Lokono Indigenous Peoples

A case is currently pending before the IACHR that, *inter alia*, explicitly challenges the establishment and management of protected areas in an indigenous territory and seeks the restitution of indigenous lands incorporated therein. It was submitted by the Kaliña and Lokono indigenous peoples of northeast Suriname and was declared admissible by the IACHR in 2007, meaning that if the facts are proven true they would likely constitute violations of the rights guaranteed by the ACHR (IACHR, 2007b). The case complains about three nature reserves within their territory: the Galibi Nature Reserve (1969), the Wane Kreek Nature Reserve (1986), and the Wia Wia Nature Reserve (1966) (VIDS, 2009). The Wane Kreek Reserve alone covers around 50 percent of the indigenous peoples' traditional territory.

These reserves were established without the Kaliña's and Lokono's participation and consent, and they negatively affect their rights on an ongoing basis. This is acknowledged in the Galibi Nature Reserve Management Plan 1992–1996, which states that "the villagers were not involved in the decision-making process. They were confronted with the reserve as a *fait accompli*" (Reichart, 1992, p30).

Suriname's 1954 Nature Protection Act makes no reference to the existence of indigenous peoples, nor does it recognize or protect their rights to their traditional territories. The same is true for Surinamese law in general (IACHR, 2006a; IACtHR, 2007). Article 1 of the act provides that the president may designate lands and waters belonging to the state as a nature reserve. As indigenous territories are legally classified as state lands, the state may unilaterally declare any indigenous territory or part thereof to be a nature reserve by decree. The act also makes no provision for the exercise of indigenous peoples' rights within nature reserves. Instead, in Article 5 of the act, hunting, fishing, or damage to the soil or the flora and fauna within the reserves are strictly prohibited and punishable as criminal offences. While this prohibition remains in force for indigenous peoples, large-scale bauxite mining, authorized by the state, is taking place in the Wane Kreek Reserve.

Applying the IACHR and the Court's jurisprudence to this situation, it is clear that the Kaliña and Lokono peoples have protected property rights to their traditional territory irrespective of whether these rights are recognized in Suriname's law. The nature and extent of these rights is defined in the first instance by the indigenous peoples' customary laws and traditional tenure systems. Should the state wish to establish or justify the existence of a protected area, it would have to show that the protected area in question complies with the applicable legal norms, including FPIC where the integrity of those territories may be affected or where the protected area may substantially interfere with culturally important economic activities (IACtHR, 2007; HRC, 2009).

Where the Kaliña and Lokono have been dispossessed of their traditional lands without their consent, and provided that they continue to maintain some degree of material or cultural/spiritual connection to the lands in question, they hold an ongoing right of restitution that is integral to satisfying their property and other rights. These lands therefore must be returned unless the state can demonstrate that there are "justifiable and concrete" reasons that prevent it from doing so. This is a requirement that will be very difficult for the state to satisfy unless the areas' protected status itself is judged to be a justifiable and concrete reason.

If we assume for the sake of argument that the IACHR or the Court will find that the protected status of land constitutes a valid excuse from the restitution requirement, application of the Court's jurisprudence should further require that the interests of the state in maintaining its proprietary rights in the protected area then be weighed against the rights and interests of the Kaliña and Lokono. In undertaking such an analysis, the Court stresses that indigenous peoples' territorial rights are fundamentally related to collective rights of survival, and that their control over territory is a necessary condition for the reproduction of culture, their development, and their ability to preserve their cultural patrimony (IACtHR 2005a, para 146). It should also be noted in this context, that the Court also has held that restricting or denying indigenous peoples access to their traditional means of subsistence is prohibited by the ACHR (IACtHR, 2005b, para 186–187).

Given indigenous peoples' fundamental and compelling interests in maintaining their relationships with their territories, the state will be hard pressed to demonstrate that its interests are paramount and should prevail. This is especially the case given the size of the protected areas (more than 50 percent of the Kaliña's and Lokono's traditional territory) and the fact that indigenous ownership *per se* does not preclude the continuation of ecosystem or species protection measures, or even the continuation of the protected areas themselves. The continuation of protected area status would nevertheless have to be negotiated and consented to by the Kaliña and Lokono.

Conclusions

The preceding issues are not new: indigenous peoples' territorial rights and restitution were extensively discussed at the 2003 World Parks Congress, and the 2003 Durban Accord: Action Plan acknowledges that there is "an urgent need to re-evaluate the wisdom and effectiveness of policies affecting indigenous peoples and local communities" (2003 Durban Accord, p25). The accord's "key targets" include full respect for the rights of indigenous peoples in relation to all existing and future protected areas; and, by 2010, the establishment and implementation of "participatory mechanisms for the restitution of indigenous peoples' traditional lands and territories that were incorporated within protected areas without their free and informed consent" (2003 Durban Accord, p26). However, there appears to be little will to meet these targets on the part of governments or conservation organizations (Colchester et al, 2008).

Irrespective, the post-Durban jurisprudence of the Court, AfCom, and others affirms and elaborates upon the nature and extent of indigenous peoples' territorial rights, FPIC, and the right to restitution. This jurisprudence also demonstrates that respect for these rights is not a matter of discretionary conservation policy and targets, but is instead a matter of international legal obligation for the countries of the Americas and the Caribbean. The 2004 decision on protected areas adopted by the CBD Conference of Parties should also be read consistently with this jurisprudence and thus provides a much needed human rights perspective to understanding international environmental law in this area. Indeed, the two bodies of law should not be viewed as mutually exclusive, but as interrelated and complementary. This will require a substantial reformulation of protected areas laws and institutions, both to remedy past violations of indigenous peoples' rights and to ensure that these rights are protected in the future.

Similarly, the Charter of the World Bank's Forest Carbon Partnership Facility affirms that its operating principles include compliance with World Bank operational policies, "taking into account the need for effective participation of forest dependent indigenous peoples and forest dwellers in decisions that may affect them, respecting their rights under national law and applicable international obligations" (FCPF, 2008, Operating Principles 3.1(d)). However, while this language is positive, there have been numerous complaints by indigenous peoples that their rights have not been adequately respected in REDD "readiness" activities supported by the FCPF to date. In the case of Indonesia, these complaints have been upheld twice by CERD under its urgent action procedure (CERD, 2009a, 2009b).

Article 10(c) of the CBD and its future elaboration in a decision of the Conference of Parties provide fertile ground to address some of the deficits in conservation practice related to indigenous peoples' rights. They also provide ample opportunity to merge environmental and human rights norms and to ensure that the protection of biological diversity and ecosystems not only takes into account the rights of indigenous peoples, but is fully consistent with those rights. This will require addressing land and resource tenure rights, recognizing indigenous peoples' rights to control and freely determine how best to utilize their territory and resources, and developing and implementing a framework for negotiating mutually acceptable and beneficial conservation agreements with indigenous peoples. By supporting this, conservation organizations and governments can demonstrate that they are serious about protecting biological diversity, combating climate change, respecting human rights, and engaging in respectful relationships with indigenous peoples.

References

AfCom (African Commission on Human Rights) (2010) *Centre for Minority Rights Development (Kenya) and Minority Rights Group International on Behalf of Endorois Welfare Council v Kenya*, Communication 276/2003, AfCom, 4 February

Amend, S. and T. Amend (eds) (1992) *Espacios sin habitantes? Parques nacionales de America del Sur*, IUCN, Gland, Switzerland

CBD (Convention on Biological Diversity) (1997) *Traditional Knowledge and Biological Diversity*, UNEP/CBD/TKBD/1/2, 18 October 1997

CBD (2004) "Decision VII/28 Protected Areas", in *Decisions Adopted by the Conference of Parties to the Convention on Biological Diversity at its Seventh Meeting*, UNEP/BDP/COP/7/21

CERD (Committee on the Elimination of Racial Discrimination) (1997) *General Recommendation XXIII (51) Concerning Indigenous Peoples*, 18 August 1997, UN Doc CERD/C/51/Misc.13/Rev.4

CERD (2001) *Concluding Observations of the Committee on the Elimination of Racial Discrimination: Sri Lanka. 14/09/2001*, UN Doc A/56/18

CERD (2002) *Concluding Observations of the Committee on the Elimination of Racial Discrimination: Botswana. 23/08/2002*, UN Doc A/57/18

CERD (2003) *Concluding Observations of the Committee on the Elimination of Racial Discrimination: Ecuador. 21/03/2003*, UN Doc CERD/C/62/CO/2

CERD (2006) *Concluding Observations of the Committee on the Elimination of Racial Discrimination: Guatemala, 15/05/06*, UN Doc CERD/C/GTM/CO/11

CERD (2007) *Concluding Observations of the Committee on the Elimination of Racial Discrimination: Ethiopia, 20/06/2007*, UN Doc CERD/C/ETH/CO/15

CERD (2008) *Concluding Observations of the Committee on the Elimination of Racial Discrimination: United States, 8 May 2008*, UN Doc CERD/C/USA/CO/6

CERD (2009a) *Communication of the Committee Adopted Pursuant to the Early Warning and Urgent Action Procedures*, 13 March 2009
CERD (2009b) *Communication of the Committee Adopted Pursuant to the Early Warning and Urgent Action Procedures*, 28 September 2009
CESCR (Committee on Economic, Social and Cultural Rights) (2001) *Concluding Observations of the Committee on Economic, Social and Cultural Rights: Colombia. 30/11/2001*, UN Doc E/C.12/Add. 1/74
Colchester, M. (2006) "Forest peoples, customary use and state forests: The case for reform", Paper presented to the 11th Biennial Congress of the International Association for the Study of Common Property (IASCP) Bali, Indonesia, 19–23 June
Colchester, M. et al (2008) *Conservation and Indigenous Peoples: Assessing Progress Since Durban*, Forest Peoples Programme, Moreton in Marsh, UK
Dooley, K., T. Griffiths, H. Leake, and S. Ozinga (2008) *Cutting Corners: World Bank's Forest and Carbon Fund Fails Forests and Peoples*, FPP-FERN briefing
FCPF (Forest Carbon Partnership Facility) (2008) *Charter of the Forest Carbon Partnership Facility*, World Bank, Washington, DC
Griffiths, T. (2008) *Seeing REDD: Forests, Climate Change Mitigation and the Rights of Indigenous Peoples and Local Communities – Update for Poznan*, Forest Peoples Programme, Moreton in Marsh, UK
HRC (Human Rights Committee) (2006) *Concluding Observations of the Human Rights Committee: Canada, 20/04/2006*, UN Doc CCPR/C/CAN/CO/5
HRC (2009) *Ángela Poma Poma v. Peru*, Communication No 1457/2006, UN Doc CCPR/C/95/D/1457/2006, 24 April 2009
IACHR (Inter-American Commission on Human Rights) (1984) *Report on the Situation of Human Rights of a Segment of the Nicaraguan Population of Miskito Origin*, OEA/Ser.L/V/II.62, Doc 26
IACHR (2002) *Report No 75/02, Mary and Carrie Dann (United States), Case No 11.140*, 27 December 2002
IACHR (2004) *Report No 40/04, Maya Indigenous Communities of the Toledo District*, Case 12.053 (Belize), 12 October 2004
IACHR (2006a) *Report on Admissibility and Merits No 09/06 on the Case of the Twelve Saramaka Clans*, 2 March 2006
IACHR (2006b) *Report No 29/06, Admissibility, Garifuna Community of "Triunfo de la Cruz" and Its Members*, 14 March 2006
IACHR (2007a) *Social Inclusion: The Road towards Strengthening Democracy in Bolivia*, OEA/Ser.L/V/II, Doc 34, 28 June 2007
IACHR (2007b) *Report No 76/07, Admissibility, The Kaliña and Lokono Peoples (Suriname)*, 15 October 2007
IACtHR (Inter-American Court of Human Rights) (2001) *Mayagna (Sumo) Awas Tingni Community Case*, 31 August, 2001, Series C No 79
IACtHR (2005a) *Yakye Axa Indigenous Community v. Paraguay*, 17 June 2005, Series C No 125
IACtHR (2005b) *Moiwana Village v. Suriname*, Judgment of 15 June 2005, Series C No 124
IACtHR (2006) *Sawhoyamaxa Indigenous Community v. Paraguay*, 29 March 2006, Series C No 146
IACtHR (2007) *Saramaka People v. Suriname*, Judgment of 28 November 2007, Series C No 172

IACtHR (2008) *Saramaka People v. Suriname. Interpretation of the Judgment on Preliminary Objections, Merits, Reparations and Costs*, Judgment of 12 August 2008, Series C No 185

IACtHR (2010) *Indigenous Community of Xákmok Kásek v. Paraguay*, Judgment of 24 August 2010, Ser C No 214

Redford, K. and E. Fearn (2007) *Protected Areas and Human Displacement: A Conservation Perspective*, Wildlife Conservation Society, Working Paper no 29, April, New York, NY

Reichart, H. A. (1992) *Galibi Natuurreservaat Beheersplan [Galibi Nature Reserve Management Plan] 1992–1996*, Paramaribo

Sobrevila, C. (2008) *The Role of Indigenous Peoples in Biodiversity Conservation: The Natural but Often Forgotten Partners*, World Bank, Washington, DC

SRRF (Special Rapporteur on the Right to Food) (2009) *See Large-Scale Land Acquisitions and Leases: A Set of Core Principles and Measures to Address the Human Rights Challenge*, SRRF, 11 June 2009

United Nations (1998) *United Nations Guiding Principles on Internal Displacement*, UN Doc E/CN.4/1998/53/Add.2, 11 February 1998

van Boven, T. (1993) *Study Concerning the Right to Restitution, Compensation and Rehabilitation for Victims of Gross Violations of Human Rights and Fundamental Freedoms*, UN Doc. E/CN.4/Sub.2/1993/8

VIDS (2009) *Securing Indigenous Peoples' Rights in Conservation in Suriname: A Review*, VIDS, October

4

Human Rights-Based Approaches to Conservation: Promise, Progress … and Pitfalls?

Jessica Campese and Grazia Borrini-Feyerabend

Conservation and sustainable use of forests and other natural resources can contribute to human rights (cf. Laban et al, 2009; Springer and Studd, 2009). Likewise, forest conservation can be enhanced where people are empowered and rights are secure (Ostrom, 1990; Molnar et al, 2004; Sunderlin et al, 2008). Conservation also protects the rights and interests of future generations, nature, and the global public. Conservation, however, can infringe on human rights through, *inter alia*, physical, economic, social, or cultural displacement (Brockington and Igoe, 2006; Redford and Fearn, 2007), or inhumane enforcement measures (cf. Jana, 2007; Anaya, 2009).

In order to address these multiple linkages and enhance human rights accountability, rights-based approaches (RBAs)[1] are emerging in forest conservation as they are in protected areas, water resources management, species conservation, climate change adaptation and mitigation, and other conservation contexts.

This chapter reviews some key rationales of RBAs to conservation; explores early progress among various actors, focusing on international conservation organizations; reflects on potential pitfalls; and offers recommendations for the way forward. We argue that while RBAs can help to reconcile human rights and forest conservation, and harness their positive synergies, achieving this, in practice, is challenging. Who should take what action? Under what circumstances? What constitutes "success", particularly in situations of competing claims over scarce natural resources? While not offering definitive answers to these questions – which generally need context-based examination and tailored solutions (cf. Chapter 8 in this volume) – the chapter proposes that RBAs may be most effective and sustainable when carried out *within appropriate governance systems*. While addressing various conservation contexts, the concepts and lessons are all applicable to forests.

Rights-Based Approaches to Conservation

Governmental, civil society, and private conservation organizations have legal, ethical, and pragmatic responsibilities to address their human rights impacts. Rights–conservation linkages are also shaped by history and their broader political, socio-economic, cultural, and ecological contexts (IUCN/CEESP, 2007, 2010; Nelson, 2010). Integrating rights thus requires looking at both conservation practice and the broader governance contexts in which it is embedded. RBAs are being promoted as one way of meeting these complex responsibilities in practice.

While there is no consensus on their definition, RBAs generally integrate rights standards, norms, and principles within conservation policy, programming, practice, and outcomes assessment. The broad aim is to ensure that conservation respects human rights in all cases, and furthers their realization wherever possible (Campese, 2009).

RBAs typically focus on human rights as recognized in international law. This includes substantive rights (e.g. life, health, food, housing, freedom to practice culture, and freedom from discrimination) and procedural rights (e.g. participation, information, access to justice, and equality before the law) commonly conceived of as minimum standards for human dignity. Included are also individual and collective rights of indigenous peoples (e.g. self-determination; free, prior, and informed consent (FPIC); control and management of lands and resources; and redress). As discussed in Chapter 1 of this volume, compared to approaches focusing exclusively on tenure, RBAs grounded in human rights are thus both broader, insofar as they encompass a wider range of rights, and narrower, insofar as they focus on minimum standards.

RBAs sometimes go beyond internationally enshrined human rights to include, for instance, customary forest rights that may not be formally recognized. According to Blomley et al (2009, p3), focusing *exclusively* on formal law can be counterproductive because such laws are not universally applied or widely understood, traditional or customary law can be more important for conflict resolution, and formal law's emphasis on state–citizen relationships can overlook other duty bearers, including the private sector and non-governmental organizations (NGOs).

RBAs complement other approaches to realizing social justice and human well-being in conservation (Springer et al, 2011). However, some approaches (e.g. community-based natural resource management) have been criticized for having weak and inconsistent impacts, particularly in cases of elite capture or difficult trade-offs with conservation objectives (Wilshusen et al, 2002; Brockington, 2003; Nelson, 2010). Forest rights approaches focused exclusively on tenure may also have insufficient benefits (Colchester, 2007). RBAs may provide a stronger foundation for integrating social considerations within forest conservation by, *inter alia*:

- recognizing vulnerable groups and individuals as rights-holders rather than 'beneficiaries';
- strengthening the accountability of duty-bearers to human rights norms and standards; and
- linking rights and responsibilities.

Rights, responsibilities, and shared resources

While grounded in human rights standards, RBAs to forest conservation raise challenging questions about various actors' rights and responsibilities (cf. Blomley et al, 2009; Springer et al, 2011). States are the primary human rights duty-bearers. Their duties include:

- *Respect*: refrain from interfering with people's pursuit or enjoyment of their rights.
- *Protect*: ensure that third parties, including businesses and NGOs, do not interfere with people's pursuit or enjoyment of their rights.
- *Promote and fulfill*: create an enabling environment for people to realize their rights, and provide more directly for the right where people are unable to do so for themselves.

The practical implications of these responsibilities are not necessarily straightforward in forest conservation. How can competing rights between forest communities be reconciled in the context of resource scarcity? How should global and intergenerational values of forests be weighed against immediate local resource needs (Blomley et al, 2009)? Are "development rights" always paramount where they conflict with conservation needs?

Non-state actors, including businesses and NGOs, are also duty-bearers, though the nature and scope of their responsibilities are debated and contested. Even when states are unable or unwilling to hold them accountable, businesses have increasingly recognized responsibilities to *respect* rights, including avoiding complicity (ICHRP, 2002; Ruggie, 2008), and arguably also have positive obligations to protect those for whom they have direct responsibility (Jungk, 2001). Some have extended these arguments to conservation NGOs, given their powers to influence and affect human rights (cf. Alcorn and Royo, 2007; Bennett and Woodman, 2010). Conservation organizations can also help further rights protection, promotion, and fulfillment.

The Conservation Initiative on Human Rights (CIHR) provides one example of how NGOs may approach their rights-related responsibilities and opportunities in forest conservation. CIHR is a consortium of international conservation organizations, including BirdLife International, Conservation International (CI), Fauna & Flora International (FFI), the IUCN (International Union for Conservation of Nature), The Nature Conservancy (TNC), Wetlands International (WI), the Wildlife Conservation Society (WCS), and WWF (World Wildlife Fund). CIHR was established in 2008 with the aim of improving the practice of conservation by "promoting the effective integration of human rights in conservation policies and practices" (Springer et al, 2010, p81). Several CIHR organizations[2] signed a framework, including commitments to:

- respect human rights;
- promote human rights within conservation programs;
- protect the vulnerable; and
- encourage good governance (CIHR, 2010).

The CIHR organizations thus recognize a baseline responsibility to refrain from carrying out or contributing to rights infringements (respect), and make additional commitments to further support and promote rights where doing so is in line with their conservation work. The *CIHR Framework* (CIHR, 2010) refers specifically to "internationally proclaimed human rights".

RBAs also raise challenges regarding rights-holders' responsibilities. Everyone has responsibilities not to infringe on the rights of others. The practical implications of this, however, may be unclear. For instance, forest communities with longstanding occupation may have strong claims to local resources; but can such claims negate the basic human rights of internally displaced people, landless people, or others who may have no other option than accessing the same resources (cf. Blomley et al, 2009)? Alternatively, can they negate the rights of future generations to enjoy them unspoiled? What is "fair" in such instances will vary with local realities. Many traditional societies have devised sophisticated systems to resolve shared or competing access and use of natural resources, with solutions usually in the hands of local institutions capable of flexible judgments. Formal-legal approaches to rights enforcement, in contrast, strive to be universal. RBAs can potentially provide a bridge between such approaches, as further explored below.

RBAs do not provide a strict formula for resolving complexities regarding rights and responsibilities in forest conservation. They can, however, provide a powerful framework for identifying key issues and negotiating competing claims. RBAs can also identify the minimum standards that should not be "traded off" (Filmer-Wilson and Anderson, 2005, p29).

The Winding Road from Theory to Practice

There has been a recent shift in RBA to conservation discourse from a strong focus on conflicts (IUCN/CEESP, 2007) to increasing discussion of opportunities and solutions (Campese et al, 2009; IUCN/CEESP, 2010). However, rights–conservation conflicts persist, RBAs to forest conservation are only emerging, and *comprehensive* examples remain scarce.

Who takes action for RBAs?

Various actors are carrying out RBAs to conservation (IUCN/CEESP, 2010). Efforts by conservation organizations are receiving substantial attention, perhaps because such organizations have also been criticized for their roles in human rights impacts (IUCN/CEESP, 2007, 2010). Their progress is laudable, though much work remains. Promising examples include NGOs working in partnership with indigenous peoples and local communities to protect customary lands and territories (Painter, 2009), and NGO-facilitated natural resource governance processes grounded in human rights (Laban et al, 2009; Springer and Studd, 2009). The IUCN Environmental Law Program has illustrated how a "step-wise" RBA might be applied in forest conservation and other contexts (Greiber et al, 2009). There are also movements towards comprehensive rights integration, such as in the above-described CIHR. NGOs that implement RBAs to *development* also often have illustrative natural resources management components (Castillo and Brouwer, 2007; Blomley et al, 2009).

Social movements, grassroots organizations, and other local people's initiatives are also demanding rights-based conservation or human rights recognition and redress in forest conservation. Many indigenous peoples and local communities are exercising their rights through continuation and revitalization of customary forest governance. Local people's efforts are often supported by civil society organizations (Cronkleton et al, 2008; IUCN/CEESP, 2010; Jonas et al, 2010).

Given the dynamic and multi-stakeholder space in which RBAs are emerging, NGOs should avoid the risk, referenced in Chapter 1, of "appropriating" rights issues. NGOs can, however, complement and support local people's efforts – for example, with regard to Indigenous Peoples' Conserved Territories and Areas Conserved by Indigenous Peoples and Local Communities (ICCAs) (IUCN, 2008a, 2008b). ICCAs have been defined as "natural and/or modified ecosystems, containing significant biodiversity values, ecological benefits and cultural values, voluntarily conserved by indigenous peoples and local communities, both sedentary and mobile, through customary laws or other effective means" (Borrini-Feyerabend et al, 2010, pp3). While recognized in the Convention on Biological Diversity (CBD) and IUCN protected areas policy, ICCAs are often not formally recognized or appropriately supported by states, and are under serious threat. Stevens (2010, p193) states that "failure to recognize ICCAs constitutes the failure to recognize many of Indigenous peoples' rights and freedoms". NGOs can help to support ICCAs through *appropriate* recognition. Recognition that is not culturally respectful, however, can undermine the institutions and relationships upon which ICCAs are built. The ICCA Consortium spells out "dos and don'ts" for outside actors willing to support ICCAs in empowering and appropriate ways (Borrini-Feyerabend et al, 2010).

Climate change mitigation through reduced emissions from deforestation and degradation and enhanced carbon stocks (REDD+)[3] also presents opportunities and challenges for conservation organizations in supporting forest communities' rights struggles (see also Chapter 5 in this volume). Numerous conservation organizations are supporting REDD+ (Poffenberger and Smith-Hanssen, 2009). At the same time, many indigenous peoples' and local communities' organizations are still developing responses to REDD+ and/or opposing it, based, in part, on rights-related risks (GFC, 2009). While conservation organizations' positions can include support for social safeguards, their promotion of REDD+ may pre-empt or undermine some indigenous peoples' and local communities' own efforts regarding rights protection in mitigation schemes. This is not to say that conservation organizations should never engage in REDD+. They should, however, refrain from supporting REDD+ carried out without indigenous peoples' and local communities' free, prior, and informed consent. They should also proceed extremely cautiously in making claims about how REDD+ relates to local people's rights where doing so may pre-empt or conflict with local people's own perspectives and actions.

What constitutes success?

While there is no consensus on the definition of "success" in RBAs to conservation, we posit that it involves comprehensively integrating rights within conservation such that rights are respected in all cases, and further realized wherever possible. To evaluate effectiveness, participatory social/rights impact assessments can be undertaken before, during, and after conservation activities, as appropriate, to help prevent and/or respond to rights concerns and opportunities. Relevant tools include those developed for protected areas (Schreckenberg et al, 2010), climate change (CCBA and CARE, 2010), and general conservation programming (Filmer-Wilson and Anderson, 2005; Svadlenak-Gomez, 2007; Greiber et al, 2009). Rights impact assessment and compliance tools developed for other sectors (e.g. Lenzen and d'Engelbronner, 2009) can also be adapted for conservation. Much work remains, however, towards development and implementation of effective tools.

Given the dynamic issues involved, "success" in RBAs may also ultimately be a factor of the *processes* used to identify and negotiate claims, and the powers shaping decisions. According to Borrini-Feyerabend et al (2004a, p12, emphasis added):

> *... a key question is how to assign responsibilities fairly and effectively – including restrictions in resource access and use – while*

> *maintaining an overall rights-based approach. The answer seems to lie in* moving away from imposed restrictions to the participatory definition of, and agreement on, shared rules.

In other words, in many cases, only a legitimate and participatory system of decision-making, assessment, and dispute resolution – grounded in rights standards and working through time – can address the complexities involved in RBAs to conservation. Such a system would effectively constitute a *rights-based system of governance* for forests or other natural resources, as further described below.

Ensuring accountability

Ensuring mutual accountability in RBAs will require further learning, but existing mechanisms can serve as models. For some examples, Rees and Vermijs (2008) analyze mechanisms used in the business arena. The non-profit organization Natural Justice (Lawyers for Communities and the Environment) and partners are supporting communities to develop Biocultural Community Protocols, a self-affirming process that enables communities to set out their own terms for the access, use, and equitable sharing of benefits of local resources and traditional knowledge, and for engaging with external actors to realize their self-determined values, priorities, and future plans (Jonas et al, 2010; www.naturaljustice.org). The NGO Survival International is promoting a draft code of conduct for some conservation activities (Bennett and Woodman, 2010).

RBAs and Governance

IUCN broadly understands governance as "the interactions among structures, processes and traditions that determine how power is exercised, how decisions are taken on issues of public concern, and how citizens or other stakeholders have their say" (Dudley, 2008, p82). For protected areas (and, by inference, forests and other natural resources), key elements include *governance type* (who possesses authority and responsibility, and is accountable for management and conservation results?) and *governance quality* (are decisions taken according to broadly accepted norms and principles?) (Borrini-Feyerabend, 2004). Governance principles discussed in the conservation context include legitimacy and voice; subsidiarity; fairness; avoidance of harm; direction; performance; accountability; transparency; and respect for human rights (Dudley, 2008, p28). Notably, these overlap substantially with the human rights principles promoted within the United Nations Common Understanding on RBAs (e.g. equity, non-discrimination, participation, inclusion, accountability, and rule of law) (UN, 2003).

Further research is needed to understand how best to support RBAs to forests. However, it is plausible that a legitimate forest governance system (appropriate *type*) adhering to sound principles (*quality*) could best facilitate an RBA by allowing rights-holders and duty-bearers to understand and negotiate the practical implications of their respective rights and responsibilities, hold one another accountable for their realization, and genuinely empower the most vulnerable. Addressing rights through such governance systems would allow for directly engaging with issues of power (authority, responsibility, and accountability). It would also move *from a snapshot approach* ("are we respecting rights with this specific decision?") *to a dynamic process* for identifying issues as they arise, negotiating claims, and taking sound decisions over time. Moreover, an effective governance system may best reconcile traditional context-based solutions with the "universal" requirements of legal systems.

The governance "type" that will best support RBAs will vary by circumstance. Extrapolating from an accepted typology for protected areas (Borrini-Feyerabend, 2004; Dudley, 2008), four broad forest governance types are *governance by government* at various levels; *shared governance* via multi-party decision-making and advisory bodies; *private governance* by landowners and concession owners; and *governance by indigenous peoples and local communities* with customary and/or legal collective rights. The legitimacy of a forest governance type, and its capacity to secure human rights, will depend on, *inter alia*, the perspectives of the actors involved and the historical processes that shaped the current arrangements. It may be useful to start from whatever form of governance is in place (or, if this is unclear, whatever form would be "legal" according to the state) and examine that arrangement in light of the aims of an RBA. Does this type of governance, in this context, allow for *respecting* rights, including those of the *most vulnerable*? Does it further *protect, promote, and fulfill* rights? Does it impede them? The answers to these questions may identify procedural and substantive rights concerns that could be solved by a change in governance type (in addition to improvements in governance *quality*). In this sense, a move towards a more appropriate governance type would constitute a move towards *improved governance*. For instance, in the case of forest governance by government or a private entity, rights advocates may argue that shared governance would better take into account local biodiversity knowledge and management skills, and better provide for the long-term food and biodiversity-based needs of the forest resident communities (Eghenter and Labo, 2003). In other circumstances, recognized *governance by indigenous peoples and local communities* may be most effective to secure rights (Stevens, 2010). In other cases, governance by government or private entities may be most appropriate.

While providing a promising focus for RBAs, having appropriate governance types in place does not necessarily *guarantee* outcomes in line with human rights standards. Thus, safeguards, assessments, and access to remedy would always be needed to ensure that the basic substantive and procedural human rights of all, and particularly those of the most vulnerable, are indeed respected, protected, and further fulfilled.

Potential Pitfalls

RBAs face several potential pitfalls (see also Laban et al, 2009). For one, RBAs may be only selectively applied. To be meaningful, such approaches should be comprehensively integrated and should involve at least respecting rights in *all* cases, including where there are conflicts between rights standards and conservation objectives. Limited RBA case examples are emerging, with several demonstrating some elements of both "synergy" and "conflict" (Campese et al, 2009; IUCN/CEESP, 2010). However, cases of successful RBAs involving the most difficult conflicts are rare. For example, under what circumstances is resettlement justified (Redford and Fearn, 2007), and what would an RBA imply? What are the "rights of nature", and how can those be reconciled when in conflict with human rights? Examples of *comprehensive* human rights integration in conservation, including in cases of difficult conflicts, are likely to emerge as practice continues.

RBAs also face significant implementation gaps. Siegele et al (2009, p69) review many conservation-relevant rights provisions in hard and soft law, but point out that good practice does not necessarily follow from good law. Reasons for this implementation gap include rights-holders' and duty-bearers' poor understanding of relevant laws (Nguyen et al, 2008); lack of technical capacity and resources to carry out known responsibilities (Stevens, 2010); lack of know-how and political will to engage in *effective* and *genuine* power-sharing and devolution of authority and benefits (Borrini-Feyerabend et al, 2004b); and lack of accountability due to, *inter alia*, inadequate governance and corruption (Nelson, 2010). These weaknesses demonstrate the importance of building rights-holders' and duty-bearers' capacity and opportunity for RBAs. They also point to a more challenging need to engage in the politics and governance systems that shape powers, rights, and responsibilities around forests and other natural resources. This is no small feat. The alternative, however, may be circumscribed RBAs that make some ground for equity, but fail to address the underlying issues shaping rights–conservation linkages.

Any actor's scope of influence will necessarily be limited, and this also poses challenges for realizing the aims of RBAs. To take one example, while conservation organizations may have substantial power (e.g. over operations in a particular site), their ability to alter broader governance systems and political forces may be weak. They will thus have to choose careful strategies to maximize their positive impacts, and mitigate against directly or indirectly supporting rights infringements. This can include seeking partnerships with human rights organizations, and/or more directly supporting local organizations. In extreme cases (e.g. where support for biodiversity conservation will further the oppression of people) (Noam, 2007), RBAs may imply that conservation organizations should "walk away" from the situation, while also taking appropriate measures to ensure that people are not made worse off by their doing so.

Care should also be exercised to ensure genuine, responsive, and empowering approaches as RBAs become "scaled out" or mainstreamed. Otherwise,

RBAs risk becoming rhetorical tools to meet external demands, rather than constituting authentic commitments. RBAs also risk becoming inflexible frameworks, rather than necessarily dynamic processes; indeed, describing the complex act of rights integration as "an RBA" may itself falsely imply an approach that can be rigidly applied (pers comm, Jenny Springer, November 2010). Similarly, RBAs should not become externally imposed or "top-down" frameworks. Rather, they should be genuinely empowering of rights-holders, and should recognize diversity in laws, values, rights, and responsibilities (Jonas et al, 2010).

RBAs also have risks. An *explicit* focus on rights is not appropriate or safe in all cases. Highlighting relative powers, claims, and duties can exacerbate underlying conflicts (Filmer-Wilson and Anderson, 2005; Blomley et al, 2009). Raising alleged violations may put claims-makers at risk of retribution. Discussions on rights should not be avoided simply because they are difficult; but they must be approached with an in-depth understanding of the context, and with respect for the safety of all those involved. This may require third-party monitoring and support.

Finally, an *exclusive* focus on rights, to the exclusion of attention to mutual responsibilities and benefits between conservation and livelihood security, can undermine opportunities for synergy (Crawhall, 2010; Jonas et al, 2010). Crawhall (2010) argues for maintaining complementary approaches, including mutual benefits between conservation objectives and local people's knowledge and capacity, rather than focusing exclusively on rights as adversarial claims.

RBAs and the IUCN

Many issues explored in this chapter are reflected in RBA development within international conservation organizations such as the IUCN. The IUCN is a union of governmental and non-governmental member organizations with the common vision of *a just world that values and conserves nature* (www.iucn.org/about). The IUCN has strong mandates to ensure just conservation processes and outcomes, including through RBAs. Members have been adopting resolutions and recommendations on equity, good governance, gender, and rights issues for over a decade (cf. Siegele et al, 2009). At the 2008 World Conservation Congress, for example, members adopted resolutions that, *inter alia*:

- request that the IUCN facilitate exchange, develop understanding and capacity, and promote and support RBAs to conservation; and call on the IUCN to develop a policy on conservation and human rights (IUCN, 2008d); and
- endorse the United Nations Declaration on the Rights of Indigenous Peoples, and instruct the IUCN "to identify and propose mechanisms to address and redress the effects of historic and current injustices against indigenous peoples in the name of conservation" (IUCN, 2008c).

Numerous other resolutions provide a mandate for the IUCN to promote appropriate governance types and quality, for protected areas, in particular, and for forests, biodiversity, and natural resources, in general. The IUCN is also part of the above-described CIHR.

The IUCN has substantial practical experience with addressing rights, including within its Forest Conservation Program.[4] A Center for International Forestry Research (CIFOR)–IUCN compilation identifies preliminary RBA lessons learned (Campese et al, 2009). The Environmental Law Center, in addition to its step-wise approach, hosts an interactive web-portal which aims to promote and encourage the adoption of RBAs to conservation and to serve as space for information-sharing (www.rights-based-approach.org). The IUCN Commission on Environmental Law and Commission on Environmental, Economic, and Social Policy have both advanced rights issues within the union (e.g. IUCN/CEESP, 2007, 2010; Greiber et al, 2009).

Undoubtedly, however, much challenging work remains to integrate rights within the IUCN in a *comprehensive* manner, and to understand the experiences and lessons that exist already. Next steps may include development of a rights policy, and supporting guidance, tools, and capacity strengthening, as well as formulation of accountability mechanisms (IUCN, 2008d; CIHR, 2010). Given the complexities and controversies involved, the IUCN is likely to face many challenges in taking these steps and avoiding the above-mentioned pitfalls.

A Way Forward?

RBAs can help reconcile forest conservation and human rights, and harness their positive synergies, and should be pursued for those reasons. Such approaches are gaining momentum among NGOs, indigenous and local communities, and other actors. Useful lessons are emerging about how RBAs can be operationalized. RBAs, however, face potential pitfalls that should be prevented. This includes the risk that they become solely rhetorical tools, or mechanisms for 'rights-washing' (pers comm, Holly Shrumm, December 2010). Genuine rights integration may require, *inter alia*, altering power relations over natural resources (e.g. through appropriate governance systems that operate through time). Ensuring respect for human rights will also mean accepting difficult trade-offs with conservation visions. RBAs also present risks, including exacerbated conflict. In sum, RBAs are not a panacea, should not exclude other equitable approaches, and should be pursued with creativity, openness, and respect for the complex realities in which conservation is often situated. What, then, are our recommendations?

We believe that conservation NGOs and other organizations should continue to document and share experiences on RBAs, including through respectful and rigorous research, and should build in-house and partner capacity. Development of practical standards, methods, and tools for assessment, communications, and accountability should also continue.

We also believe, however, that furthering RBAs to forest conservation – by NGOs, communities, governments, civil society, and others – may best be served by supporting the development of governance systems grounded in clear recognition of the claims, rights, and responsibilities of all relevant actors. Such systems should include rights-based assessment and access to remedy. Avenues should be provided for rights-holders, particularly the most vulnerable, to engage in their own informed analysis of the issues and options at stake, and to take a meaningful role in planning, decision-making, implementation, and learning by doing. Through such processes, some form of shared governance will likely be developed, although the possibility of actors successfully arguing for self-determination should also be supported. In all cases, RBAs should remain dynamic and responsive. Through such processes, competing claims can be understood and negotiated, and rights–conservation synergies – whatever forms they may take – will have a better chance to flourish.

Acknowledgements

The views expressed in this chapter are those of the authors, and do not necessarily reflect those of the IUCN or any other organization. The authors thank Louisa Denier, Holly Shrumm, and Jenny Springer for their helpful comments. Sections of the chapter also build upon Campese (2009).

Notes

1. There is no strict distinction in the literature between RBAs and "human rights-based approaches". While speaking of an approach grounded in human rights, we refer to "RBAs" to be consistent with common usage and to maintain open dialogue regarding which rights and frameworks should be considered.
2. The common *CIHR Framework* has been approved by CI, FFI, IUCN, WI, WWF, and TNC. WCS and BirdLife have adopted principles and measures based on the common framework, adapted to their own organizational contexts.
3. The "+" indicates inclusion of the role of conservation, sustainable management of forests, and enhancement of forest carbon stocks.
4. For details on the work of individual IUCN programs, see www.iucn.org.

References

Alcorn, J. and A. G. Royo (2007) "Conservation's engagement with human rights: 'Traction', 'slippage', or avoidance?", *Policy Matters*, vol 15, pp115–139

Anaya, J. S. (2009) *Report by the Special Rapporteur on the Situation of Human Rights and Fundamental Freedoms of Indigenous Peoples, James Anaya. Addendum: Report on the Situation of Indigenous Peoples in Nepal*, UN Human Rights Council, A/HRC/12/34/Add.3, 20 July

Bennett, G. and J. Woodman (2010) "Code of conduct for conservation and indigenous peoples: An initial proposal", *Policy Matters*, vol 17, pp84–92

Blomley, T., P. Franks, and M. R. Maharian (2009) *From Needs to Rights: Lessons Learned from the Application of Rights Based Approaches to Natural Resource Governance in Ghana, Uganda and Nepal*, Rights and Resources Initiative, Washington, DC

Borrini-Feyerabend, G. (2004) "Governance of protected areas, participation and equity", in Secretariat of the Convention on Biological Diversity (ed) *Biodiversity Issues for Consideration in the Planning, Establishment and Management of Protected Areas Sites and Networks*, Montreal, Canada, CBD Technical Series no 15, pp100–105

Borrini-Feyerabend, G., A. Kothari, and G. Oviedo (2004a) *Indigenous and Local Communities and Protected Areas: Towards Equity and Enhanced Conservation*, IUCN, Gland, Switzerland, and Cambridge, UK

Borrini-Feyerabend, G., M. Pimbert, T. Farvar, A. Kothari, and Y. Renard (2004b) *Sharing Power: Learning by Doing in Co-Management of Natural Resources throughout the World*, IIED and CEESP/IUCN, London

Borrini-Feyerabend, G. et al (2010) *Bio-Cultural Diversity Conserved by Indigenous Peoples and Local Communities: Examples and Analysis*, ICCA Consortium and CENESTA for GEF SGP, GTZ, IIED and IUCN/CEESP, Tehran, Iran

Brockington, D. (2003) "Injustice and conservation: Is local support necessary for sustainable protected areas?", *Policy Matters*, vol 12, pp22–30

Brockington, D. and J. Igoe (2006) "Eviction for conservation: A global overview", *Conservation and Society*, vol 4, no 3, pp424–470

Campese, J. (2009) "Rights-based approaches to conservation: An overview of concepts and questions", in J. Campese, T. Sunderland, T. Greiber, and G. Oviedo (eds) *Rights-Based Approaches: Exploring Issues and Opportunities for Conservation*, CIFOR and IUCN, Bogor, Indonesia

Campese, J., T. Sunderland, T. Greiber, and G. Oviedo (eds) (2009) *Rights-Based Approaches: Exploring Issues and Opportunities for Conservation*, CIFOR and IUCN, Bogor, Indonesia

Castillo, G. E. and M. Brouwer (2007) "Reflections on integrating a rights based approach in environment and development", *Policy Matters*, vol 15, pp153–167

CCBA (Climate, Community and Biodiversity Alliance) and CARE International (2010) *REDD+ Social and Environmental Standards*, Version 1, June

CIHR (Conservation Initiative on Human Rights) (2010) *Conservation and Human Rights Framework*, September, http://cmsdata.iucn.org/downloads/cihr_framework_e_sept2010_1.pdf

Colchester, M. (2007) "Beyond tenure: Rights-based approaches to peoples and forests. Some lessons from the Forest Peoples Programme", Paper presented to the International Conference on Poverty Reduction in Forests: Tenure, Markets and Policy Reforms, Forest Peoples Programme, Bangkok, Thailand, 3–7 September 2007

Crawhall, N. (2010) "Valuing traditional ecological knowledge: Supplementing a rights-based approach to sustainability in Africa", *Policy Matters*, vol 17, pp227–231

Cronkleton, P., P. L. Taylor, D. Barry, S. Stone-Jovicich, and M. Schmink (2008) *Environmental Governance and the Emergence of Forest-Based Social Movements*, Occasional Paper 49, CIFOR, Bogor, Indonesia

Dudley, N. (2008) *Guidelines for Applying Protected Area Management Categories*, IUCN, Gland, Switzerland

Eghenter, C. and M. Labo (2003) "The Dayak people and Kayan Mentarang – first co-managed protected area in Indonesia", *Policy Matters*, vol 12, pp248–253

Filmer-Wilson, E. and M. Anderson (2005) "Integrating human rights into energy and environment programming: A reference paper", Prepared for UNDP
GFC (Global Forest Coalition) (2009) *REDD Realities: How Strategies to Reduce Emissions from Deforestation and Forest Degradation Could Impact on Biodiversity and Indigenous Peoples in Developing Countries*, GFC, Asunción, Paraguay, December
Greiber, T., M. Janki, M. Orellana, A. Savaresi-Hartmann, and D. Shelton (2009) *Conservation with Justice: A Rights-Based Approach*, IUCN Environmental Policy and Law Paper 71, IUCN, Gland, Switzerland
ICHRP (International Council on Human Rights Policy) (2002) *Beyond Voluntarism: Human Rights and the Developing International Legal Obligations of Companies*, ICHRP, Versoix, Switzerland
IUCN (International Union for Conservation of Nature) (2008a) *IUCN Resolution 4.049: Supporting Indigenous Conservation Territories and other Indigenous Peoples' and Community Conserved Areas*, Adopted at the 4th Session of the World Conservation Congress, October 2008
IUCN (2008b) IUCN *Resolution 4.050: Recognition of Indigenous Conservation Territories*, Adopted at the 4th Session of the World Conservation Congress, October 2008
IUCN (2008c) *IUCN Resolution 4.052: Implementing the United Nations Declaration on the Rights of Indigenous Peoples*, Adopted at the 4th Session of the World Conservation Congress, October 2008
IUCN (2008d) *IUCN Resolution 4.056: Rights-based approaches to conservation*, Adopted at the 4th Session of the World Conservation Congress, October 2008
IUCN/CEESP (2007) "Conservation and human rights", *Policy Matters*, vol 15, pp1–370
IUCN/CEESP (2010) "Exploring the right to diversity in conservation law, policy, and practice", *Policy Matters*, vol 17, pp1–251
Jana, S. (2007) "Voices from the margins: Human rights crises around protected areas in Nepal", *Policy Matters*, vol 15, pp87–100
Jonas, H., H. Shrumm, and K. Bavikatte (2010) "Biocultural community protocols and conservation pluralism", *Policy Matters*, vol 17, pp106–116
Jungk, M. (2001) *Defining the Scope of Business Responsibility for Human Rights Abroad*, Human Rights and Business Project, Danish Centre for Human Rights, Confederation of Danish Industries, Industrialization Fund for Developing Countries, Copenhagen, Denmark
Laban, P., F. Haddad, and B. Mizyed (2009) "Enhancing rights and local level accountability in water management in the Middle East: Conceptual framework and case studies from Palestine and Jordan", in J. Campese, T. Sunderland, T. Greiber, and G. Oviedo (eds) *Rights-Based Approaches: Exploring Issues and Opportunities for Conservation*, CIFOR and IUCN, Bogor, Indonesia, pp97–122
Lenzen, O. and M. d'Engelbronner (2009) *Guide to Corporate Human Rights Impact Assessment Tools*, Aim for Human Rights, The Netherlands
Molnar, A., S. J. Scherr, and A. Khare (2004) *Who Conserves the World Forests? Community Driven Strategies to Protect Forests and Respect Rights*, Forest Trends and Agricultural Partners, Washington, DC
Nelson, F. (ed) (2010) *Community Rights, Conservation and Contested Land: The Politics of Natural Resource Governance in Africa*, Earthscan, London

Nguyen, Q. T., N. Ba Ngai, T. Ngoc Thanh, W. Sunderlin, and Y. Yasmi (2008) *Forest Tenure Reform in Vietnam: Case Studies from the Northern Upland and Central Highlands Regions*, RECOFTC and RRI, Bangkok, Thailand, and Washington, DC

Noam, Z. (2007) "Eco-authoritarian conservation and ethnic conflict in Burma ", *Policy Matters*, vol 15, pp272–287

Ostrom, E. (1990) *Governing the Commons: The Evolution of Institutions for Collective Action*, Cambridge University Press, Cambridge, UK

Painter, M. (2009) "Rights-based conservation and the quality of life of indigenous people in the Bolivian Chaco", in J. Campese, T. Sunderland, T. Greiber, and G. Oviedo (eds) *Rights-Based Approaches: Exploring Issues and Opportunities for Conservation*, CIFOR and IUCN, Bogor, Indonesia, pp164–185

Poffenberger, M. and K. Smith-Hanssen (2009) "Forest communities and REDD climate initiatives", *Asia Pacific Issues*, vol 91, pp1–8, www.communityforestry international.org/publications/research_reports/REDD_Climate_Initiatives2009.pdf

Redford, K. H. and E. Fearn (2007) *Protected Areas and Human Displacement: A Conservation Perspective*, World Conservation Society, Working Paper no 29, Bronx, NY

Rees, C. and D. Vermijs (2008) *Mapping Grievance Mechanisms in the Business and Human Rights Arena*, Corporate Social Responsibility Initiative Report no 28, Harvard Kennedy School, Cambridge, MA

Ruggie, J. (2008) *Protect, Respect and Remedy: A Framework for Business and Human Rights – Report of the Special Representative of the Secretary-General on the Issue of Human Rights and Transnational Corporations and Other Business Enterprises*, UN Human Rights Council, A/HRC/8/5, 7 April

Schreckenberg, K., I. Camargo, K. Withnall, C. Corrigan, P. Franks, D. Roe, L. M. Scherl, and V. Richardson (2010) *Social Assessment of Conservation Initiatives: A Review of Rapid Methodologies*, Natural Resource Issues no 22, International Institute for Environment and Development (IIED), London

Siegele, L., D. Roe, A. Giuliani, and N. Winer (2009) "Conservation and human rights: Who says what?", in J. Campese, T. Sunderland, T. Greiber, and G. Oviedo (eds) *Rights-Based Approaches: Exploring Issues and Opportunities for Conservation*, CIFOR and IUCN, Bogor, Indonesia, pp47–76

Springer, J. and K. Studd (2009) "The conversatorio for citizen action: Fulfilling rights and responsibilities in natural resource management in Colombia", in J. Campese, T. Sunderland, T. Greiber, and G. Oviedo (eds) *Rights-Based Approaches: Exploring Issues and Opportunities for Conservation*, CIFOR and IUCN, Bogor, Indonesia, pp77–96

Springer, J., J. Gastelumendi, G. Oviedo, K. Walker Painemilla, M. Painter, K. Seesink, H. Schneider, and D. Thomas (2010) "The Conservation Initiative on Human Rights: Promoting increased integration of human rights in conservation", *Policy Matters*, vol 17, pp81–83

Springer, J. and J. Campese with M. Painter (2011) *Scoping Paper on Key Issues at the Intersection of Conservation and Human Rights*, Prepared for CIHR

Stevens, S. (2010) "Implementing the UN Declaration on the Rights of Indigenous Peoples and International Human Rights Law through the recognition of ICCAs", *Policy Matters*, vol 17, pp180–194

Sunderlin, W., J. Hatcher, and M. Liddle (2008) *From Exclusion to Ownership? Challenges and Opportunities in Advancing Forest Tenure Reform*, Rights and Resources Initiative, Washington, DC

Svadlenak-Gomez, K. (2007) *Human Rights and Conservation: Integrating Human Rights in Conservation Programming*, TransLinks Paper no 48, World Conservation Society (WCS), Bronx, NY

UN (United Nations) (2003) *UN Common Understanding on the Human Rights Based Approach to Development Cooperation*, Developed at the Inter-Agency Workshop on a Human Rights-Based Approach in the Context of UN Reform, 3–5 May 2003

Wilshusen, P., S. Brechin, C. Fortwangler, and P. West (2002) "Reinventing a square wheel: Critique of a resurgent 'protection paradigm' in international biodiversity conservation", *Society and Natural Resources*, vol 15, pp17–40

Part II

What Claims Find Support?

It is a fairly straightforward thing to call for the recognition of forest people's claims as rights, whether at the global or other levels. It is a lot more difficult to determine what kinds of claims find support, and what specific content forest rights entail. This is not just a question about the different categories of rights at the heart of the three distinct origins of the forest rights agenda developed in Part I (i.e. tenure rights, rights of self-determination, and human rights). The issue is that defining forest people's claims gets even more difficult when one considers their concrete meaning. The demand for tenure rights, for example, remains an abstract notion that requires concretization in specific contexts. Tenure rights may or may not involve ownership, and ownership may take on very different concrete forms in different countries or villages.

Insights from agricultural land reform demonstrate the pitfalls of simplistic definitions of claims and their consequent universalization as rights. For a long time, advocates of land reform considered ownership the most appropriate – and strongest – tenure right to transfer to local people. Much effort went into programs either bestowing ownership on new land or formalizing people's existing tenure rights as ownership. Yet, the outcomes of these programs were often disappointing. Some of the new landowners lost their rights as quickly as they had gained them. Some of the existing rights-holders did not see their rights strengthened, but had to deal with an additional source of competition and insecurity. In agriculture, therefore, rights activists have learned that there are no universal silver bullets for the recognition of land claims as rights.

This question about the kinds of claims is the central concern in the next three chapters. The chapters thereby take up the first critical debate among forest rights activists that we identified in Chapter 1: what claims find the support of forest rights activists? The chapters all grapple with the underlying tension between the single notion of a right in the abstract and the many ways in which to tangibly interpret such a right.

Ribot and Larson deal with another important debate about the nature of forest rights in Chapter 5: do changes in administrative rights and regulations empower forest people in the absence of broader political reforms? Drawing on a case study of charcoal production in Senegal, they examine the role of political economy in determining whether and how different groups are able to access new opportunities afforded by regulatory reforms. They find that implementation of forestry reforms on uneven political and economic playing fields further excludes the poor despite the stated intention to enhance their inclusion. Actual policy implementation aggravates social and economic imbalances – allowing the wealthy to extract further benefit from desired natural resource markets. Narrow regulatory reforms inadvertently compound existing impoverishment and limit marginalized groups' access to political, social, and economic processes, unless the broader differences between poor and rich are addressed. Forest rights cannot be realized through minor changes in existing regulatory regimes, even where the discourse and aims surrounding policy reforms may be valid and sincere.

Barnhart indicates in Chapter 6 how the recognition of a narrow set of rights can support the articulation of broader rights demands. State recognition of tenure rights in Nepal, she finds, facilitates demands among forest user groups for a broader spectrum of rights – including social, economic, and political rights. The state grants local groups community forest management powers. The groups granted tenure rights, in turn, are able to establish social and political rights, in independent mobilization as well as with support from relevant partner organizations. Under certain preconditions, forest tenure rights thus can provide a platform for further recognition of basic human rights.

In Chapter 7, To Xuan Phuc's discussion of forest rights in Vietnam offers a different perspective on the question of whether the recognition of a very specific set of rights serves to empower forest people by comparing tenure rights with economic rights. He analyses households' abilities to derive income from forestland holdings recently allocated by the state. Households cannot derive economic benefits from forest holdings without access to complementary productive resources and labor power. Subsequently, poor villagers are often forced to sell their land holdings to better-off villagers and outsiders. Poverty conditions are also reinforced by the scarcity of government credit and expansion of commodity production. Already advantaged households consequently accumulate further benefits from devolution, whereas poor households sell their land and/or incur debts. These differentiating effects of tenure transfers can only be avoided if legal transfers are accompanied by suitable measures to level broader economic and political inequalities (i.e. efforts to strengthen economic rights).

The three chapters thus indicate the difficulties of defining forest people's claims and suitable rights in a universal manner. This note of caution also emerges from other analyses in the book, such as Dorondel's observation that villagers can only take full advantage of their tenure rights if they can hold local government accountable (see Chapter 12). Forest people assert a large repertoire of claims, and their claims find the support of forest rights activists in different forms. The concrete rights asserted and supported, in practice, include rights to specific forest resources, rights to forests seen as bundles of multiple resources, rights to sell forest products, rights to hold local government accountable, rights to legal tenure, rights to economic benefits, human rights, individual rights, collective rights, etc. This multiplicity of rights poses a vexing challenge to forest rights activists, as there are many ways to translate abstract rights into operational rights.

The underlying issue is that forest people's claims and the articulation of rights is dynamic and context specific. Rights are historically contingent, as they reflect the particular histories of claims, struggles for rights, past incidences of recognition or lack thereof, etc. Rights are also specific to particular contexts, as they refer to concrete forest landscapes, cultural meanings, and ongoing forms of political contestation. Rights, in other words, are nothing "natural" to be either recognized or denied, and their realization depends on wider political economic dynamics. Actual rights are the result of concrete practices and broader political economic processes taking place in particular cultural and ecological settings.

At the same time, the insights generated in this and other parts of the book demonstrate the significance of broader political and economic rights. Recognition of a narrow set of forest rights – even ownership – is unlikely to enable forest people's empowerment unless forest people are already in a position of relative power. Regulatory reforms in the forestry sector have to come along with broader political and economic change to overcome the entrenched political economic inequalities characterizing forest areas across the world. Broader political reforms are required to level the uneven playing field and to enable marginalized people to mobilize their legal forest rights, a message that also emerges very strongly in Chapters 9, 10, 12, and 15. Economic empowerment is a precondition for disadvantaged forest people to benefit from tenure transfers.

5

Affirmative Policy on an Uneven Playing Field: Implications for REDD

Jesse C. Ribot and Anne M. Larson

Reduced emissions from deforestation and degradation and enhanced carbon stocks (REDD+)[1] is a global program for disbursing funds, primarily to pay national governments in developing countries to reduce forest carbon emission (UN-REDD, 2009, p4). Its framers acknowledge that REDD risks "decoupling conservation from development", enabling "powerful REDD consortia to deprive communities of their legitimate land-development aspirations", undermining "hard-fought gains in forest management practices", and eroding "culturally rooted not-for-profit conservation values" (FAO et al, 2008, pp4–5).[2]

The framers view these risks as balanced by ecological sustainability, "the potential to achieve significant sustainable development benefits for millions of people worldwide", and to "help sustain or improve livelihoods and food security for local communities". In addition, they foresee that "a premium may be negotiable for emission reductions that generate additional benefits" for local people. They even acknowledge "that REDD benefits in some circumstances may have to be traded off against other social, economic or environmental benefits" and call for care in taking local place-based complexity into account when designing REDD interventions (FAO et al, 2008, pp4–5).

What will prevent the promised "premium" from being competed down to nothing, as is the tendency in any competitive market (Economics 101)? Who will do the trading-off of REDD benefits? Isn't the converse – local needs being traded off for REDD carbon benefits – more likely? These trade-offs involve people's lives and histories at the edge of the legal world. How will REDD proponents ensure that trade-offs are just? How will REDD strategies take their needs and aspirations into account? How will rights be established and enforced?

Safeguards are, of course, being developed. The Center for International Forestry Research (CIFOR) and others have called for REDD+ to systematically address effectiveness, efficiency, equity, and co-benefits – what they call the three Es+ (3Es+) (Angelsen et al, 2009). The UN-REDD Program and others call for legal instruments and stipulations to protect local forest-based communities, such as a right to free, prior, informed consent (FPIC), in a global convention or national legislation to protect indigenous forest people (Colchester, 2010; UN-REDD, 2010). A Norwegian government report (Angelsen et al, 2009) proposes developing principles to promote participation: "Definition of rights to lands, territories, and resources, including ecosystem services; representation in REDD decision making, both internationally and nationally, including access to dispute resolution mechanisms; and integration of REDD into long-term development processes".

The proposed principles are all excellent. Nevertheless, their application has been tried many times, and the results have been less than stellar (Lemos and Agrawal, 2006; Tacconi et al, 2006; Lund et al, 2009; Larson et al, 2010; Ribot et al, 2010). The forestry and conservation institutions that are asked to apply them resist being subjected to such principles. The complexity of an illegible (*à la* Scott, 1998) context also makes implementation very difficult. Most programs and associated protections have not addressed the needs and aspirations, or established and protected the rights, of resource-dependent rural populations. Rural people remain seriously unrepresented as well as under-represented in forestry matters (Ribot et al, 2010).

Legal reforms of statutory rights are only one set of instruments and factors shaping access to forests and forest benefits. They are important; but their creation, application, effectiveness, and ultimate meanings are shaped by entrenched rural inequalities embedded in disabling social, political-economic, and legal hierarchies. Lack of empowered representation along with policy-backed marginalization are deepened even by so-called "neutral" or seemingly "fair" policies because of unequal access to capital, labor, and credit, rooted in class, identity, and social relations (Baviskar, 2001; Ribot and Peluso, 2003; Larson et al, 2006; Bandiaky, 2007). Together these factors slant the access playing field, pitting marginal people against the more powerful, reshaping the intention and effects of legal instruments.

Legal reforms are easily fettered, stymied, manipulated, and circumvented. Local people are often given strong rights to valueless resources, rights to forests rather than markets, rights to implement rather than decide, and rights to participate rather than control. In this context, will informed consent be selectively limited to questions involving the inadequate "rights" held by the rural poor? Will it, too, be extracted, coerced, cajoled, persuaded, or hoodwinked out of communities? Will it remain selectively targeted to indigenous peoples, leaving out the many non-indigenous longstanding forest communities who deserve equal protection? Real enfranchisement and emancipation require the establishment of universal representation – via empowered and locally accountable authorities. This remains the central challenge to fair and just REDD+.

There must be feedback mechanisms so that societies can react to and adjust as laws are made, implemented, and repurposed, and as practices (regulated or not) shape their lives. Stratification is a constant process. Hence, we need constant counter-processes to hold decision-makers publicly accountable. Change takes place through iterative processes linking legal instruments with discursive/social/political-economic context. It will not be enough to tweak and enforce existing 'rights' – especially since the rights worth having are usually held only by the rich – a product of failed representation and long histories of extractive and market-oriented regimes. Rights to markets and lucrative resources cannot continue to be reserved for elites while the poor are relegated to labor opportunities and use rights. Positive change will require a radical rethinking – indeed, dismantling – of forestry regulation and management in addition to establishing and strengthening of substantive rights and representation for forest-based people.

This chapter takes an "access" approach to policy analysis, described below, by analyzing the political economy that shapes the distribution of benefits from forests under a particular policy regime. It focuses on the real-world problem that forest policies and/or policy implementation systematically exclude various groups from forest benefits. In doing so, forestry policies and practices, sometimes inadvertently, impoverish and maintain the poverty of these groups. Poverty is not just about being left out of economic growth. It is produced by the very policies that enable some to profit: today from timber, firewood and charcoal; tomorrow from carbon.

The remainder of this chapter is organized into three sections. The first frames our access approach. The second presents a case study of charcoal production in Senegal – little to do with REDD yet, but everything to do with the uneven fields on which REDD is already beginning to play out. The third is a synthesis and conclusion.[3]

From Disabling to Enabling Policies: Rights with Access

Governments have long mediated forest access (Thompson, 1977; Scott, 1998). Sunderlin et al (2005, p1390) describe how "forestry laws and regulations in many countries were written to [ensure] privileged access to timber wealth and to prevent counter-appropriation by the poor". In Africa, the colonial antecedents of many of today's forestry policies were unapologetic in favoring Europeans over Africans (Ribot, 1999a). Writing on Gabon, for example, the colonial historian R. L. Buell reported that:

> *... before 1924, natives held [forest] concessions and sold wood upon the same basis as Europeans. But the competition became so keen ... that in a 1924 administrative order, the government declared that a native could not cut and sell wood except for his own use without making a deposit with the government of twenty-five hundred francs – a prohibitive sum.* (Buell, 1928, vol II, p256)

Over 80 percent of the world's forests are on public lands, and the state is often the first gateway to forest access (FAO, 2006).[4] Forestry authorities are still using many exclusionary strategies directly descendent from these earlier techniques, keeping forest peoples poor.

The World Bank (2002) estimates that 1.6 billion people depend on forests for livelihoods (see also Kaimowitz, 2003). At least in some countries, there is an important correlation between forests and poverty (Blaikie, 1985; Peluso, 1992; Dasgupta, 1993; Taylor et al, 2006). Communities living in and near forests suffer from outsiders' commercial exploitation of forest resources (see Colchester et al, 2006a, for a list of studies and consequences; Ribot, 2004; Oyono et al, 2006), and it is clear from commodity chain and forest-village studies that vast profits are extracted through many commercial forest activities, yet little remains local (Blaikie, 1985; Peluso, 1992; Dasgupta, 1993; Ribot, 1998, 2006). Retaining forest benefits locally may offer options for improved well-being in these areas. Indeed, the great commercial and subsistence value of forests is drawing increased attention to their potential role in poverty alleviation (Kaimowitz and Ribot, 2002; Oksanen et al, 2003; Sunderlin et al, 2005), though there may also be trade-offs between forest conservation and poverty alleviation (Wunder, 2001; Tacconi et al, 2006; Lund et al, 2009).

Over the past two decades there has been a wave of reforms designed to increase local participation and benefits for forest-dwellers. Studies of community forestry in Mexican *ejidos* (Bray, 2005) and Guatemala's Petén (Gómez and Mendez, 2005; Taylor, 2006) have demonstrated substantial economic and other livelihood benefits, such as increased income, greater human and social capital, natural resource conservation, decreased vulnerability, greater equity, democratization of power, and empowerment. Community forestry in Cameroon and Nepal has also significantly increased income to forest villages (Agrawal, 2001, 2005; Oyono, 2004, 2006). But few such studies are available precisely because communities rarely have policy-supported access to forests, the resources that are valuable in them, or policy-supported access to the capital and markets that would make increased income possible (Ribot, 1998, 2004). These experiments in inclusion are important trail-blazers towards more progressive and pro-poor forestry; but they still represent only small enclaves of change in the vast wilderness of forestry practice.[5]

Important efforts to solve problems in the forest sector have focused on illegal logging, while including concerns about the rights of forest-based populations – such as the World Bank-supported Forest Law Enforcement and Governance

(FLEG) process (World Bank, 2006). This attention to illegal logging, however, is predicated on two questionable assumptions: first, that "illegal is unsustainable" and "legal is sustainable" (Colchester et al, 2006b), and, second, that the illegal is merely a matter of disrespecting laws that are otherwise appropriate. Legal forestry and forestry laws, however, are not always based on criteria of sustainability. If diligently followed, many regulations would not result in sustainable management (Ribot, 1999a, 2006). Furthermore, forestry laws define the boundaries of the legal – a domain of 'legal' that may not be realistic or just. Since forestry laws discriminate against small and collective forest land and resource users – often banning their access to necessary goods – these users are driven to illegal practices.

The FLEG process comes at these issues from a different perspective. The World Bank (2006) emphasizes stopping forest crime, identifying poverty as one of its drivers. Hence, reforming land tenure and biased regulations that produce poverty is necessary to "help address the poverty-related driver" (World Bank, 2006, pxi). Therefore, the World Bank emphasizes that explicitly addressing the ensemble of means by which these groups are excluded and by supporting inclusion may also help to reduce the illegal logging that it views as a cause of deforestation. As Colchester et al (2006b) argue, FLEG initiatives should address all the laws affecting forest-dependent peoples (not just forestry laws), adopt a rights-based approach, and be linked to governance reform processes that promote broad-based participation, accountability, and transparency in natural resource management. Reforming forestry laws is not enough – constitutions, organic codes, laws of decentralization, electoral codes, tax laws, fiscal codes, laws establishing rights of assembly, and co-operative laws are all also implicated in the powers, rights, and representation of forest villagers (Ribot, 2004).

Colchester et al (2006a) point out that many governments have signed numerous "soft laws" such as international agreements that, among other things, recognize indigenous land rights and customary resource management practices, but that these have rarely been incorporated within forestry legislation. In cases where land rights have been granted, this does not necessarily include rights over trees or forest management.[6] Where laws have been passed granting communities greater access to land and/or forests, these have often been adopted through processes outside the realm of forest policy specifically, such as in Nicaragua's autonomous regions or Panama's indigenous *comarcas*, though there are exceptions, such as Bolivia (Larson et al, 2006). For their part, forest policy frameworks tend to be developed with the significant influence of timber interests, as well as state and multilateral financial institutions, but less often, despite the widespread discourse, with the effective participation of community or indigenous groups (Silva et al, 2002). It is no surprise that forest policy usually reflects multiple interests – at the expense of these under-represented forest-dependent actors.

How do we explain the paradox of increasing recognition of *rights* on a broad scale alongside the failure to guarantee basic access in practice? The rights-based approach to livelihoods emphasizes the importance of grounding development in human rights legislation, based on international norms and laws. It is attempting to re-politicize development and bring in normative, pragmatic, and ethical issues by empowering people to make claims against their governments and demand accountability (cf. Ferguson, 1996; Nyamu-Musembi and Cornwall, 2004). But how are such rights to be translated into practice? Why is it that legislating new rights rarely translates into greater benefits for average rural citizens?

In their "theory of access", Ribot (1998) and Ribot and Peluso (2003) contrast the common formulation of property as a "bundle of rights", with their conception of access as a "bundle of powers". To gain access to forest resources, guaranteed property rights – temporarily, such as short- or long-term contracts for concessions, or permanently, such as land titles or constitutional guarantees[7] – are a necessary first step; but the power to act on those rights depends on the negotiation of a number of complementary access mechanisms. The access approach highlights the role of power, emphasizing that many people gain and maintain access *through others* who control it. Thus, on state forest lands, it is usually the central forestry authority that determines who has (legal) access rights to the forest, and on these as well as private, including collective, forest-lands, it is the central forestry authority that determines who will have access to permits for the (legal) use and/or sale of forest resources. In the cases we present below, regulations and the authorities who implement and enforce them systematically favor logging companies and create multilayered access barriers for communities and smallholders – even when those communities and smallholders hold secure rights to the forest resource itself.

The access approach complements the rights-based approach. Rights-based approaches, if practiced according to their original conception, aim to alter power dynamics in development (Nyamu-Musembi and Cornwall, 2004). In this framework, gaining rights, such as those established through the signing of international treaties and inscribed in national laws, is only a first step. Rights, however, only take effect when implemented in practice – also a political process that will likely challenge vested interests at every step. On the ground, then, a rights-based approach is successful when the power dynamics of access are altered and access to livelihood assets are improved for formerly excluded and marginalized groups.

The case below shows how current forestry policies in Senegal – even when called community based or participatory – and the ways in which they are selectively implemented continue to reproduce the double standards and conditions that disadvantage, create, and maintain the rural poor.

Charcoal in Senegal

> *There is a certain complicity of the Forest Service – it is not against us, it is for the interest of the patrons.* (Elected rural council president in discussion at Tamba Atelier, with four rural council presidents, 14 February 2006)

Until 1998 the system of forest management in Senegal was organized around a system of licenses, permits, and quotas allocated by the national Forest Service. A national quota for charcoal production was fixed by the Forest Service each year. Forest Service officials and agents claimed this quota was based on estimates of the total national demand for charcoal and the potential for the forests to meet this demand. But these estimates were neither based on surveys of consumption nor forest inventories. Indeed, there was (and still is) a persistent gap between the quantity set for the quota and the much higher figures from consumption surveys. In practice, the quota is based on the previous year's quota, which is lowered or raised depending on various political considerations. Over the past decade, the quota was lowered almost every year – regardless of demand – thus increasing illegal production (since demand was always met) (Ribot, 2006).

Prior to the new decentralized forestry laws, the nationally set quota was divided among some 120 to 170 forestry *patrons*, or merchants, at the head of forestry enterprises – co-operatives, economic interest groups (GIEs), and corporations – who hold professional forest producer licenses delivered by the Forest Service. Allocation of quotas among these entities was based on their previous year's quota with adjustments based on whether or not the enterprise had fully exploited its quota and had engaged in positive forest management activities, such as reforestation. Some forestry patrons did plant trees by the side of the road to demonstrate such efforts – they called these plantations their '*chogo goro*', or bribes – since these helped them get larger quota allocations from the Forest Service. During this period, new professional licenses were also allocated most years (enabling new co-operatives to enter the market).

Each year after the allocation of quotas, the Forest Service and Ministry of Environment held a national meeting to "announce" the opening of the new season. They passed a decree listing the quotas for each enterprise and indicating in which of the two production regions, Tambacounda or Kolda, these quotas were to be exploited. Soon after, the Regional Forest Services then called a meeting in each regional capital to inform the recipients of the location they would be given to exploit their quotas. Sites were chosen by foresters based on "eyeballing" of standing wood. The forest agents organized the zone into very loose rotations and chose sites by eye, such that some areas that were considered exhausted would be closed, while others that had not been official production sites for a time would be reopened. There was no local say in the matter.

Progressive legal changes gave the rural populations new rights during the late 1990s. Senegal's 1996 decentralization law gave rural communities (the most local level of local government) jurisdiction over forests in their territorial boundaries. The rural council (the elected body governing the rural community) was given jurisdiction over "management of forests on the basis of a management plan approved by the competent state authority" (RdS, 1996a, Article 30), and the 1998 Forestry Code (RdS, 1998) gave the council the right to determine who can produce in these forests (Article L8, R21). Furthermore, even the more general decentralization framing law gave the council jurisdiction over "the organization of exploitation of all gathered plant products and the cutting of wood" (RdS, 1996b, Article 195). Finally, the Forestry Code states that "Community Forests are those forests situated outside of the forested domain of the State and included within the administrative boundaries of the Rural Community who is the manager" (RdS, 1998, Article R9). The forested domain of the state consists of areas reserved for special uses and protection (RdS, 1998, Article R2), and most of Senegal's forests are not reserved. In short, under the new laws, most rural communities control large portions of the forests – if not all of the forests – within their territorial boundaries.

To protect the rights over these forests, the Forestry Code requires the Forest Service to obtain the signature of the rural council president, elected from among the rural councilors, before any commercial production can take place in their forests (Article L4). For their part, rural council presidents (PCRs) play an executive role and cannot take action prior to a meeting and deliberation of the council whose decisions are taken by a majority vote (RdS, 1996b, Articles 200, 212). In short, the new laws require a majority vote of the rural council approving production before anyone can produce in rural community forests.

The radical new 1998 Forestry Code changed everything – at least on paper. The amount of production would be based on the biological potential of each rural community's forests rather than by decree in Dakar and the regional capital. The enterprises to work in a given forest would be chosen by the rural council rather than the National Forest Service in Dakar. If implemented, the new system would empower rural councilors to manage their forests for the benefit of the rural community. The law stated that the quota system was to be entirely eliminated in 2001 (RdS, 1998, Article R66). But despite all the new rural community rights, as of 2009, little had changed. The Forest Service continued to manage and allocate access to the forests via centrally allocated licenses, quotas, and permits.

In implementation, the rural council's new rights to decide over forest use are being attenuated by double standards concerning forest access and market access. The new laws give the rural council president rights over forests, but the Forest Service refuses to transfer the powers. Rural populations in Senegal lose out mainly due to two double standards: access to forests and access to commercial opportunities are both skewed against them. These are discussed below.

Double standards in forest access

The rural council president legally controls the rights to access forests, but foresters do not allow him to exercise his prerogative. Foresters argue that villagers and councilors are ignorant of forest management and that national priorities trump local ones. They treat the PCR's signature as a requirement rather than as a transfer of powers or change in practice. They and the merchants coerce – threaten and pressure – the council presidents to sign away forest rights (Ribot, 2008).[8] Rural council presidents say no, but are ultimately pressured to sign.

The regional Forest Service deputy director was asked: "Given that the majority of rural council presidents do not want production in the forests of their rural communities, how do you choose their rural community as a production site?" He replied with a non-comprehending look on his face: "If the PCRs have acceptable reasons, if the local population would not like ... ?" He then stated: "The resource is for the entire country. To not use it, there must be technical reasons. The populations are there to manage. There is a national imperative. There are preoccupations of the state. This can't work if the populations pose problems for development." Nevertheless, the deputy director knew the letter of the law that he was breaking every day. When asked to explain the function of the PCR's signature, he replied, "The PCR signature must come before the quota is allocated, before the regional council determines which zones are open to exploitation" (interview, deputy director of the regional Forest Service, Tambacounda, 3 December 2005). In short, rural councils are asked for their signature, but are not allowed to say no – despite the fact that the population whom they represent opposes production.

In four rural communities, where donors have set up model forest management projects, the new forestry laws are being applied – albeit selectively. In project areas, rural people have the opportunity to participate in forest exploitation, but only if they engage in forest management activities required by the Forest Service. The ecological evidence indicates that few measures are necessary since natural regeneration in the zone is robust (Ribot, 1999b). Forest villagers know this and do not see the need for most management activities. Nevertheless, to be allowed to manage their own forests, rural communities must use management plans created by the Forest Service. That is, whereas urban-based merchants install migrant laborers in non-project areas without management plans, villagers wishing to engage in charcoal production must do so under strictly supervised and highly managed circumstances (ironically, even in these areas, most of the PCRs and councilors did not want production, but were forced to sign off under pressure from the Forest Service – similarly to PCRs in non-project areas).

By creating a spatially limited implementation zone for existing policies, the projects serve as an excuse not to implement the laws more generally. Foresters argue that the projects represent cutting-edge practices that are being tested before expanding to other sites; but this argument does not justify prohibiting forest villagers outside of the production areas from producing charcoal while allocating their forests to the migrant woodcutters of the urban-based merchants. In fact, the project areas serve as a decoy. When donors come to visit the forests, they are shown project areas where management – rather, the labor to implement management obligations imposed by the Forest Service – is decentralized. They do not see the rest of the forests where Forest Service activities have barely changed since colonial times (including those areas where production is closed without the consultation of rural councils). The project, in this case, reduces the progressive 1998 forestry laws to a territorially limited experiment.

Double standards in market access

The Forest Service requires all those wishing to trade in the charcoal market (called charcoal patrons) to be members of a registered co-operative, economic interest group (GIE) or a private enterprise in order to request from the Forest Service a license (*Cart Professionnelle d'Exploitant Forestière*) in the name of their organization (see Bâ, 2006). Despite the elimination of the quota in 2001, production and marketing remain impossible without quotas, since at least until 2009 permits are still only allocated to those with quotas.[9]

Upon receipt of a professional card, the member's organization can be allocated a portion of the national quota in the annual process of quota allocation. In 2004, the national quota of 500,000 quintals was divided into 462,650 quintals of initial quotas and 37,350 quintals of encouragement quotas (7.5 percent) (RdS, 2004, pp11–12). The initial quotas are allocated at the beginning of the season and the encouragement quotas are allocated at the discretion of the Forest Service and minister later in the season (Bâ, 2006).

Each year new co-operatives and GIEs (a kind of for-profit collective business) have been added to the market. In 2005 there were 164 organizations (RdS, 2005), up by 18 new organizations from 147 organizations in 2004 (RdS, 2004, p12). Unfortunately, all of the rural-based co-operatives that we have spoken with who have requested professional cards have been refused. The quota per patron, however, is shrinking, and many patrons believe that new licenses are being allocated to relatives of powerful merchants and political allies: "The registration of new entities is due to the officials: the president of the national union and the state. Most of the entities are family businesses – brothers and sisters." In particular, they are the brothers and sisters of other already registered patrons. According to older patrons, some of the new organizations do nothing but resell their quotas to others (patron 2, 25 December 2005). As one patron told us in disgust, "Most of the large quota people are new entrants into the market" (interview, co-operative president, Patron Charbonnier, Tamba, 26 December 2005).

During recent years, the Forest Service, upon recommendation by the director of the national union, has been allocating licenses and quotas to women (interview, union leader, 22 February 2006). This is a new phenomenon. In an interview with one such woman, we learned that she was the wife of an established patron. Forming her own co-operative appears to be a strategy to increase her husband's quota (interview by Salieu Core Diallo, February 2006). Other patrons are not happy with this. One told us that the national union president "was given a supplementary quota [officially called an encouragement quota]. They give quotas and supplementary quotas to women. These women are behind the national union president" (interview, PCR workshop, 14 February 2006).

Over the past several years, rural councilors and other rural community members have requested licenses so that they can receive quotas.[10] In one case, a rural GIE president went to the director of the Forestry Service in Dakar to request the card. He explained:

> *We put together a GIE in 1998 with its own forest production unit. We filed our registration papers at Tamba [the regional capital] – it went all the way to Dakar. I saw the dossier at Hann [national Forestry Office] ... We asked for co-operative member cards and for a quota. We were discouraged. We went to Hann and to Tamba. In Dakar, they wanted to give us quotas as individuals. I said "no" in solidarity with the rest of my colleagues with whom I was putting together the GIE.* (Interview, elected rural council member, Tambacounda Region, 22 December 2005)

A similar story was recounted by a GIE president in Missirah (interview, December 2005).

The Forestry Service explains its refusal to give professional cards to local GIEs by saying "they need to be trained" and explaining that "if we let them produce, they will learn the bad techniques of the *surga* [migrant woodcutters]" who work for the current patrons (interviews, two Inspecteur Régionale des Eaux et Forets (IREF) and three Agent Technique des Eaux et Forets (ATEF) officials in Tamba, December 2005). First, the community has to be organized into village committees and trained to manage and survey forest rotations and to use the Casamance kiln (these are all requirements within project areas, but not requirements under the law). Meanwhile, however, the Forest Service continues to admit new co-operatives that have no knowledge of production whatsoever and to hand out quotas to patrons who are producing without any training or management within managed and non-managed zones.

After the initial and encouragement quotas are allocated, illegal production and transport fill in the gap between legal supply and actual consumption. But these illegal activities can only be done by those who hold licenses and quotas – since license- and quota-holders can use their licenses to obtain supplementary permits and can hide extra charcoal with their legal loads. This is how the gap between the quota and consumption is filled. The market – legal and illegal – is tied up in the hands of a small privileged group of well-connected patrons (Ribot, 2006).

Despite the fact that Senegal's progressive forestry policies have given away little of the state's control, they are at this moment being reformed and replaced by less-progressive new forestry laws (Ribot, 2009). Senegal's current Forestry Bill takes back many of the rights hard won over the decades of decentralization. Senegal's Forest Service went from being a civilian service to a military service in 2008. No donors in forestry made any protests. This militarization contrasts with the movement in most countries. At the time of writing, the 2011 Forestry Bill is likely to soon pass. It promises to consolidate control over commercial access to forests with the Forest Service, something the current laws had threatened but never achieved. If the bill passes, then the quota will be renamed "the contract" and will have the same function as before, but under a new name (Faye and Ribot, 2010).

Conclusions

In Senegal, the forestry laws are beautifully written. They place key decisions over forest exploitation in the hands of democratic local authorities and open the markets for communities to sell their products. But these laws are not respected in practice. Old forestry laws favoring the urban elite have been eliminated by new progressive laws; but in reality, little has changed. Through long-abrogated but still-practiced policies, Senegal's Forest Service allocates licenses and quotas in order to retain market access in elite hands – licenses continue to be allocated while quotas have merely been renamed "contracts". Despite the new laws, and community demands and protests, new decentralized forestry rights and opportunities have not reached rural communities. Senegal's forest access and management standards are singular and fair in law, but conflict in practice. Urban elites are systematically favored, while rural forest-dwelling populations are excluded, with total disregard for their rights, wishes, and needs.

Inequalities favoring outside commercial interests over those of local communities are maintained in Senegal and elsewhere by a large repertoire of access means (see Ribot and Oyono, 2005; Toni, 2006; Smith, 2006; Larson and Ribot, 2008; Nayak and Berkes, 2008; He, 2010; Neimark, 2010; Saito-Jensen et al, 2010). Although the specific dynamics vary from country to country, poor communities and smallholders remain at a disadvantage in comparison to more powerful outside interests. Laws may create uniform standards or access asymmetries; they may even transfer decision-making powers and lucrative opportunities to poor rural populations. But even when laws create fair access, they are not just when unevenly implemented or selectively enforced, and they are not sufficient to overcome existing inequities unless they are designed and implemented with an affirmative approach (Ribot, 2004; Bandiaky, 2007; Baviskar, 2007).

Despite a new language concerning decentralization and the recognition of indigenous or rural peoples' rights, forest services around the world still treat local people as subjects and continue to colonize forested territories. The policies applied today are almost all – even when given a participatory or decentralized patina – relics of colonial management based on earlier European practice (as in Africa), or of post-colonial entrenched bureaucracies (as in Latin America). REDD will build on this tradition of domination if it does not seek to transform the structure and cultures of forestry and forest services. Weak checks, balances, and protections are not enough. Targeting the poor is not enough. New progressive policies will have to target the rich to shoot down some of their inordinate privilege. New policies that favor benefits for local people over outsiders are needed. New politics that regulate through minimum standards rather than maximum control may have transformative power. The poor must be represented in the making and implementation of these processes – proportionally to their inordinate numbers.

The outcomes of forest policy and implementation processes, worldwide, demonstrate the multiple and competing interests and goals of different stakeholders and the weaker power of those who lose out. The existence of apparently fair laws, however, also demonstrates that advocacy by and for forest-based populations has in some cases been successful and that further progress is possible. Senegal's forestry policies are much better for rural people today than 20 years ago. New policies should include deepening forestry decentralization through effective representation and participation, seeking common ground across myriad local goals and interests, and identifying opportunities to challenge unjust privilege. Representation will mean that when local people say "no" to the exploitation of local forests, then there will be *no* exploitation. It may not mean that when they say "yes", exploitation should necessarily take place. Such a "yes" could have negative ecological externalities for higher scales of social, economic, and political organization. Environmental standards are needed (Ribot, 2004). The right to say "no" to exploitation, however, gives them the ability to negotiate – the cost of this negotiation, the costs of real "participation" and "representation", may have to be less privilege to outside interests.

REDD is entering this slanted world with the primary objective of carbon emissions reduction – not justice or equity. If community rights are already limited, as in Senegal, will they be limited in the future under REDD in the name of carbon sequestration? What rules for resource use will be developed to meet carbon targets under REDD, who will create and enforce these rules, and how might they limit community access to forests for livelihoods? If communities carry new burdens – such as limitations on activities permitted in forests ("no" imposed from above) – will they be fairly compensated? Will the rights to forest benefits – this time to carbon funds – once again be captured by outsiders (Larson, in press)?

To improve access to benefits from forests for rural residents – whether for livelihoods, logging, or REDD funding – rights-based approaches to livelihoods must challenge power relations by transforming access. Of course, the rich and powerful have little interest in giving up their wealth and power. Rights are only real when they are enforced – rights are 'enforceable claims' (MacPherson, 1989) – so rules not enforced are not rights. The weak must have the means of enforcement, whether through representation, resistance, or withdrawal, to fight for good policies and fair implementation. The support of good analysis and of sympathetic allies (a role of scholars) can back progressive claims and help to exert pressure on those who resist change. But this is only one small contribution to reform.

Policies are damped out in the transition from discourse to law and transformed in implementation. Hildyard et al (2001) observe that participatory projects and policies "however carefully prepared, generally flounder the moment they leave the drawing board. By the time they are implemented, they are frequently unrecognizable even to their authors." Lele (2000, cited in Nayak and Berkes, 2008, p707) postulates "that (a) participatory management involves the devolution of power, (b) but the state is by nature interested in maintaining and accumulating power, and therefore (c) joint forest management must be a 'sleight of hand' carried out by state to co-opt activists and placate donors while retaining control and even expanding it in new ways".

These are fair observations; but policy is not something that is made and implemented once and for all. It is an iterative process that requires constant vigilance and struggle. Stratification is a constant process. Inequity always comes back. Governments perform (enact, portray, pretend) change while maintaining business as usual. Still, progressive policies are better than regressive ones. There are many politicians, foresters, donors, NGOs, and administrators fighting for greater justice in forestry. Their efforts can make things better even if they do not make things well. REDD will have to be hyper-progressive and affirmative if it is to benefit the rural poor.

Notes

1 The plus sign indicates inclusion of forest restoration, rehabilitation, sustainable management, and/or afforestation and reforestation.
2 For further discussion of the risks that REDD programs pose to local livelihoods, see Phelps et al (2010).
3 The case study and general framing in this chapter are based on Larson and Ribot (2008).
4 FAO (2006) reports that 84 percent of forests were publicly owned in 2000. Another study found that in developing countries, 71 percent were owned and administered by governments, and 8 percent were publicly owned but reserved for communities (White and Martin, 2002). Only in Central America are private forests (at 56 percent) more important than public (FAO, 2006).
5 Many forestry "projects" claim to increase local income. This chapter does not draw on the literature on projects – projects are not state law or policy.
6 In Colombia, Peru, and Venezuela, the state still apparently granted concessions to third parties on indigenous and community lands as of 2006 (Taylor et al, 2006).
7 Such as for indigenous communities and *quilombos* (colonies formed by runaway slaves) in Brazil (Taylor et al, 2006).
8 The story of this coercion is told in the films *Weex Dunx and the Quota and Semmiñ Ñaari Boor* (see http://doublebladedaxe.com).
9 Like the quota, the license too is illegal under Senegal's current laws (see RdS, 1995).
10 "The PCRs organized to demand their own quotas. Patron X was our point man. E and F said no, because decentralization is for protecting the forests, not to exploit them" (interview, president of Union Nationale des Coopératives des Exploitants Forestiers du Sénégal (UNCEFS), 9 July 2004).

References

Agrawal, A. (2001) "The regulatory community: Decentralization and the environment in the *Van Panchayats* (Forest Councils) of Kumaon", *Mountain Research and Development*, vol 21, pp208–211

Agrawal, A. (2005) *Environmentality: Technologies of Government and the Making of Subjects*, Duke University Press, Durham, UK

Angelsen, A., S. Brown, C. Loisel, L. Peskett, C. Streck, and D. Zarin (2009) *Reducing Emissions from Deforestation and Forest Degradation (REDD): An Options Assessment Report*, Prepared for the Government of Norway by Meridian Institute, Norway

Bâ, E. D. (2006) "Le quota est mort, vive le quota! Ou les vicisitudes de la réglementation de l'exploitation du charbon de bois au Senegal", Environmental Governance in Africa, Working Paper no 19, World Resources Institute, Washington, DC

Bandiaky, S. (2007) "Engendering exclusion in Senegal's democratic decentralization: Subordinating women through participatory natural resource management", Representation, Equity and Environment Working Paper Series, World Resources Institute, Washington, DC

Baviskar, A. (2001) "Forest management as political practice: Indian experiences with the accommodation of multiple interests", *International Journal of Agricultural Resources, Governance and Ecology*, vol 1, nos 3–4, pp243–263

Baviskar, A. (2007) "Between micro-politics and administrative imperatives: Decentralisation and the Watershed Mission in Madhya Pradesh, India", in J. Ribot and A. Larson (eds) *Democratic Decentralisation through a Natural Resource Lens*, Routledge, London, pp26–40

Blaikie, P. (1985) *The Political Economy of Soil Erosion*, Longman, London

Bray, D. (2005) "Community forestry in Mexico: Twenty lessons learned and four future pathways", in D. Bray, L. Merino-Pérez, and D. Barry (eds) *The Community Forests of Mexico: Managing for Sustainable Landscapes*, University of Texas Press, Austin, TX

Buell, R. L. (1928) *The Native Problem in Africa*, vols I and II, MacMillan Company, New York, NY

Colchester, M. (2010) *Free, Prior and Informed Consent: Making FPIC Work for Forests and Peoples*, Scoping paper prepared for The Forest Dialogue's (TFD) FPIC Initiative, Research Paper no 11, July 2010, Yale, New Haven, CT

Colchester, M. et al (2006a) "Forest peoples, customary use and state forests: The case for reform", Paper presented at the 11th Biennial Congress of the International Association for the Study of Common Property, Bali, 19–22 June

Colchester, M. et al (2006b) *Justice in the Forest: Rural Livelihoods and Forest Law Enforcement*, CIFOR, Bogor

Dasgupta, P. (1993) *An Inquiry into Well-Being and Destitution*, Oxford University Press, Oxford, UK

FAO (Food and Agriculture Organization of the United Nations) (2006) *Global Forest Resources Assessment 2005*, FAO, Rome

FAO, UNDP and UNEP (2008) *UN Collaborative Program on Reducing Emissions from Deforestation and Forest Degradation in Developing Countries (UN-REDD)*, FAO, UNDP, UNEP Framework Document, Rome

Faye, P. and J. C. Ribot (2010) *Semmiñ Ñaari Boor* [*Double-Bladed Axe*], www.doublebladedaxe.com

Ferguson, J. (1996) "Transnational topographies of power: Beyond 'the state' and 'civil society' in the study of African politics", Mimeo

Gómez, I. and V. E. Mendez (2005) *Asociación de Comunidades Forestales de Petén, Guatemala: Contexto, logros y desafíos*, PRISMA, San Salvador

He, J. (2010) "Globalised forest-products: Commodification of the matsutake mushroom in Tibetan villages, Yunnan, southwest China", *International Forestry Review*, vol 12, no 1, pp27–37

Hildyard, N., H. Pandurang, P. Wolvekamp, and R. Somesekhare (2001) "Pluralism, participation, and power: Joint forest management in India", in B. Cooke and U. Kothari (eds) *Participation: The New Tyranny*, Zed Books, London, pp56–71

Kaimowitz, D. (2003) "Not by bread alone ... forests and rural livelihoods in sub-Saharan Africa", in T. Oksanen, B. Pajari, and T. Tuomasjukka (eds) *Forests in Poverty Reduction Strategies: Capturing the Potential*, European Forest Institute Proceedings no 47, European Forest Institute, Finland

Kaimowitz, D. and J. C. Ribot (2002) "Services and infrastructure versus natural resource management: Building a base for democratic decentralization", Paper presented to the International Conference on Decentralization and the Environment, Bellagio, 18–22 February

Larson, A. (in press) "Forest tenure reform in the age of climate change: Lessons for REDD+", *Global Environmental Change*, doi:10.1016/j.gloenvcha.2010.11.008

Larson, A. and J. Ribot (2008) *Democratic Decentralisation through a Natural Resource Lens*, Routledge, London

Larson, A., P. Pacheco, F. Toni, and M. Vallejo (2006) *Exclusión e inclusión en la forestería latinoamericana: hacia donde va la descentralización?*, CIFOR/IDRC, La Paz, Bolivia

Larson, A., D. Barry, G. R. Dahal, and C. J. P. Colfer (2010) *Forests for People: Community Rights and Forest Tenure Reform*, Earthscan, London

Lemos, M. and A. Agrawal (2006) "A greener revolution in the making? Environmental governance in the 21st century", Environment, vol 49, pp36–45

Lund, J. F., K. Balooni, and T. Casse (2009) "Change we can believe in? Reviewing studies on conservation impact of popular participation in forest management", *Conservation and Society*, vol 7, no 2, pp71–83

MacPherson, C. B. (1989) "The meaning of property", in C. B. MacPherson (ed) *Property: Mainstream and Critical Positions*, University of Toronto Press, Toronto, Canada

Nayak, P. and F. Berkes (2008) "Politics of co-optation: Community forest management versus joint forest management in Orissa, India", *Environmental Management*, vol 41, no 5, pp707–718

Ndoye, O. and J. C. Tieguhong (2004) "Forest resources and rural livelihoods: The conflict between timber and non-timber forest products in the Congo Basin", *Scandinavian Journal of Forest Research*, vol 19, pp36–44

Neimark, B. (2010) "Subverting regulatory protection of 'natural commodities': The *Prunus africana* in Madagascar", *Development and Change*, vol 41, no 5, pp929–954

Nyamu-Musembi, C. and A. Cornwall (2004) *What Is the "Rights-Based Approach" About? Perspectives from International Development Agencies*, IDS Working Paper 234, IDS, Sussex, UK

Oksanen, T., B. Pajari and T. Tuomasjukka (eds) (2003) *Forests in Poverty Reduction Strategies: Capturing the Potential*, European Forest Institute Proceedings no 47, European Forest Institute, Finland

Oyono, P. R. (2004) "Institutional deficit, representation, and decentralized forest management in Cameroon", *Environmental Governance in Africa Working Paper*, no 15, World Resources Institute, Washington, DC

Oyono, P. R. (2006) "Acteurs locaux, representation et 'politics' des éco-pouvoirs dans le Cameroun rural post-1994", *Canadian Journal of Development Studies*, vol 27, pp163–185

Oyono, P. R., J. Ribot, and A. Larson (2006) "Green and black gold in rural Cameroon: Natural resources for local governance, justice and sustainability", *Environmental Governance in Africa*, Working Paper no 22, World Resources Institute, Washington, DC

Peluso, N. L. (1992) *Rich Forests, Poor People: Resource Control and Resistance in Java*, University of California Press, Berkeley, CA

Phelps, J., E. L. Webb, and A. Agrawal (2010) "Does REDD+ threaten to recentralize forest governance?", *Science*, vol 328, 16 April, pp312–313

RdS (République du Sénégal) (1995) *Décret 95 – 132 du 1 février 1995 portant libéralisation de l'accès à certaines professions*, République du Sénégal, Ministère de l'Environnement et de la Protection de la Nature, Sénégal

RdS (1996a) *Loi portant transfert de compétences aux régions aux communes et aux communautés rurales*, République du Sénégal, Dakar, 22 March

RdS (1996b) *Loi Portant Code des Collectivités Locales*, République du Sénégal, Dakar, 22 March

RdS (1998) *Code Forestier, Loi no 98/03 du 08 janvier 1998 et Décret no 98/164 du 20 février 1998*, République du Sénégal, Ministère de l'Environnement et de la Protection de la Nature, Direction des Eaux, Forets, Chasse et de la Conservation des Sols, Sénégal

RdS (2004) *Arrêté fixant les modalités d'organisation de la compagne d'exploitation forestière 2004*, Convocation no 000550/MEA/DEFCCS, Ministère de l'Environnement et de l'Assainissement, République du Sénégal, Sénégal, 8 February 2004

RdS (2005) *Arrêté fixant les modalités d'organisation de la campagne d'exploitation forestière 2005*, République du Sénégal, Ministère de l'Environnement et de la Protection de la Nature, Sénégal

Ribot, J. C. (1998) "Theorizing access: Forest profits along Senegal's charcoal commodity chain", *Development and Change*, vol 29, pp307–341

Ribot, J. C. (1999a) "Decentralization, participation and accountability in Sahelian forestry: Legal instruments of political-administrative control", *Africa*, vol 69, no 1, pp23–65

Ribot, J. C. (1999b) "A history of fear: Imagining deforestation in the West African dryland forests", *Global Ecology and Biogeography*, vol 8, no 3–4, pp291–300

Ribot, J. C. (2004) *Waiting for Democracy: The Politics of Choice in Natural Resource Decentralization*, World Resources Institute, Washington, DC

Ribot, J. C. (2006) *Analyse de la filière charbon de bois au Sénégal: Recommandations*, Policy brief based on the Senegal Dutch research program: Pour une gestion décentralisée et démocratique des ressources forestières au Sénégal, 1 September

Ribot, J.C. (2008) "Authority over forests: Negotiating democratic decentralization in Senegal", Representation, Equity and Environment Working Paper Series (formerly Environmental Governance in Africa Working Paper Series) Paper No. 36, http://pdf.wri.org/wp36_ribot.pdf (available in French at http://pdf.wri.org/ribot_french_wp36.pdf)

Ribot, J. C. (2009) *Analysis of Senegal's Draft Forestry Code: With Special Attention to Its Support for Decentralization Laws*, Report to US Agency for International Development, Senegal, 22 October, p59

Ribot, J. C. and R. Oyono (2005) "The politics of decentralization", in B. Wisner, C. Toulmin, and R. Chitiga (eds) *Toward a New Map of Africa*, Earthscan, London

Ribot, J. C. and N. L. Peluso (2003) "A theory of access", *Rural Sociology*, vol 68, pp153–181

Ribot, J. C., T. Treue, and J. F. Lund (2010) "Democratic decentralization in sub-Saharan Africa: Its contribution to forest management, livelihoods, and enfranchisement", *Environmental Conservation*, vol 37, no 1, pp35–44

Saito-Jensen, M., I. Nathan, and T. Treue (2010) "Beyond elite capture? Community based natural resource management in Mahmad Nagar village, Andhra Pradesh, India", *Environmental Conservation*, vol 37, no 03, pp327–335

Scott, J. (1998) *Seeing Like a State: How Certain Schemes to Improve the Human Condition Have Failed*, Yale University Press, New Haven, CT

Silva, E., D. Kaimowitz, A., Bojanic, F., Ekoko, T., Manurung, and I. Pavez, (2002) "Making the law of the jungle: The reform of forest legislation in Bolivia, Cameroon, Costa Rica, and Indonesia", *Global Environmental Politics*, vol 2, pp63–97

Smith, W. (2006) "Regulating timber commodity chains: Timber commodity chains linking Cameroon and Europe", Paper presented at the 11th Biennial Congress of the International Association for the Study of Common Property, Bali, 19–22 June

Sunderlin, W., A. Angelsen, B. Belcher, P. Burgers, R. Nasi, L. Santoso, and S. Wunder (2005) "Livelihoods, forests, and conservation in developing countries: An overview", *World Development*, vol 33, pp1381–1402

Tacconi, L., Y. Siagian, and R. Syam (2006) "On the theory of decentralization, forests and livelihoods", *Environmental Management and Development Occasional Papers*, No 09, Asia Pacific School of Economics and Governments, Australian National University, Canberra

Taylor, P. (2006) *Conservation, Community and Culture? New Directions for Community Forest Concessions in the Maya Biosphere Reserve of Guatemala*, Draft report for the Association of Forest Communities of Petén, Guatemala

Taylor, P., A. Larson, and S. Stone (2006) *Forest Tenure and Poverty in Latin America: A Preliminary Scoping Exercise*, Report prepared for the Center for International Forestry Research, Bogor

Thompson, E. P. (1977) *Whigs and Hunters: The Origin of the Black Act*, Penguin, Harmondsworth, UK

Toni, F. (2006) *Gestao florestal na Amazonia Brasileira*, CIFOR/IDRC, La Paz, Bolivia

UN-REDD (2009) *Rules of Procedure and Operational Guidance*, United Nations REDD Program, New York, NY

UN-REDD (2010) *UN-REDD Programme Newsletter*, no 3, October 2010, http://webcache.googleusercontent.com/search?q=cache:35tEThq81ckJ:www.un-redd.org/Newsletter13/Panama_FPIC_Workshop/tabid/6407/Default.aspx+Free+Prior+Informed+Consent+(FPIC)+REDD&cd=2&hl=en&ct=clnk, accessed December 2010

White, A. and A. Martin (2002) *Who Owns the World's Forests?*, Forest Trends, Washington, DC

World Bank (2002) *A Revised Forest Strategy for the World Bank Group*, World Bank, Washington, DC

World Bank (2006) *Strengthening Forest Law Enforcement and Governance: Assessing a Systemic Constraint to Sustainable Development*, Report no 36638GLB, World Bank, Washington, DC

Wunder, S. (2001) "Poverty alleviation and tropical forests – what scope for synergies?", *World Development*, vol 29, pp1817–1833

6

Advancing Human Rights through Community Forestry in Nepal

Shaunna Barnhart

Community forestry in Nepal is not new; its roots extend to late 1970s donor projects to promote "*panchayat* forests", which gave a limited number of local government bodies more control of forest resources. Nepal's community forestry as we now know it stems from the 1993 Forest Act and 1995 Forest Regulations, drafted under the new and fledgling democracy that flourished after a people's movement swept Nepal in the spring of 1990, seemingly sweeping away decades of authoritarian rule. Any analyst of Nepali politics will tell you that this is a naive and euphoric account, and the story is certainly not this simplistic; but it is necessary to place this government action in context, at a time when Nepalis were finding power in their collective voices.

Nepal is home to over 14,000 community forest user groups (CFUGs) controlling an estimated 23 percent of forestland. In this model, as outlined in the national forest legislation, an elected committee manages the forest resource and its benefits for communal good according to their own rules and action plan. This devolves previous government forest control to smaller localized scales of governance. But community forestry as practiced in Nepal is about more than managing resource access; it is a forum for advocating and advancing tangible human rights. The rights agenda in forestry finds its roots in social movements for redistribution of forest tenure, rights to self-determination, and human rights. These three discourses also converge in the emergence of community forestry in Nepal and in the wider circle of rights that such groups now claim. Nepal's community forestry movement now represents and enacts a broader set of rights, leveraging community forestry as an effective strategy to realize broader rights claims. CFUGs can be both advocates and practitioners of a social justice that believes in equitable access to forest benefits while providing necessary support to live life with dignity by ensuring basic human rights services such as health, education, and livelihood. Such groups have the potential to ameliorate inequality and suffering through distribution of forest benefits – benefits that include a suite of services that the community forest group can provide, resulting from their collective rights to productive forestlands.

This chapter explores the experiences of five CFUGs in Nepal's Jhapa District, whose programs advance human rights. Recognizing that the community forest experience varies greatly across groups and across Nepal (indeed, with over 14,000 CFUGs, how can it not), this chapter argues that, based on these five groups' experiences, community forestry not only provides an avenue to securing forest rights, but once those rights are codified, can then become a primary agent in securing basic human rights – including social, political, and economic rights – and creating inclusive futures through social justice, both independently and in conjunction with partner organizations.

Applying Human Rights

Economic, social, cultural, civil, and political rights are cornerstones to the modern conceptualization of human rights, or those rights needed to live a life with dignity. In order to achieve this basic right of living a dignified life, "all basic necessities of life – work, food, housing, health care, education and culture" must be attainable for all (OHCHR, 1991). These rights have been codified by the United Nations (UN) in the Universal Declaration of Human Rights (1948), the International Covenant on Economic, Social, and Cultural Rights (1966) and the International Covenant on Civil and Political Rights (1966), which focus on the role of states in guaranteeing and protecting citizens' human rights through state mechanisms. As Johnson and Forsyth (2002, p1592) summarize, rights are generally "understood as a claim to a benefit ... that states or other forms of authority have agreed to uphold", where rights are defined by a universal system "in which minimum standards of well-being are extended to the widest possible constituency".

The idea of "universal" human rights is contested. Human rights change and evolve over time, and relegating them to a static non-changing list drafted in specific historical junctures, under equally specific political and social conditions, can create a monolithic, inelastic, and non-dynamic structure in a dynamic, changing world. As Arzabe (2001, p31) argues:

> *Individual and social needs change over time, as do the rights that ensure the possibility of* being *human with dignity. The human*

> *rights system evolves, adding new rights that formerly were not considered necessary for the correct development of each one's personality in society.*

Others argue that the human rights discourse stems from a Western-paternalistic worldview where universal rights, "as defined through largely Western experiences, limits the relevance of rights to local circumstances and thereby effects yet another form of Eurocentric epistemological violence which seeks to normalize a particular and self-servicing social vision" (Mohan and Holland, 2001, p193).

Sen (2000) argues that "universal" human rights are justifiable because values termed "human rights" have a long history and documentation in non-Western cultures. The difficulty is in actually providing all with equal access to those rights and protection from their abuses. To do so requires multiple scales of governance, rendering human rights an ideal rather than a practice (Sen, 2000). This does not mean that we should abandon human rights as ineffectual and unobtainable. Instead, it becomes an issue of scale; localized efforts can be more effective in securing basic human rights than international bodies and agreements (Mertus, 2009).

Scale must be considered in understanding how human rights move from idealized discourse to everyday practice. Human rights are (largely) enshrined in international bodies, (supposedly) protected by national governments, and (often) brought into practice by organizations and groups acting in communities (Mertus, 2009). Conflict, cultural norms, social exclusions, political uncertainty, and poverty separate individuals from those rights and create barriers to accessing those rights which allow one to fulfill his/her potential and live a dignified life. In such situations, how do communities advance equity in access to basic human rights?

A rights-based approach to development is one strategy to ensure basic human rights for a wider constituency (Mohan and Holland, 2001; Sengupta et al, 2005; Aaronson and Zimmerman, 2006), as is a rights-based approach to natural resources. As Campese and Borrini-Feyerabend argue in Chapter 4, human rights are protected and advanced through sustainable forest use and preservation of natural resources. Yet, development initiatives are generally spearheaded by international aid organizations that can ultimately serve to limit the efficacy of the rights-based agenda itself (Mohan and Holland, 2001). The case presented here does not fit smoothly into this paradigm. Rather, the advancing of a broader suite of rights through established forest rights groups is based upon a grassroots demand; forest users are identifying their own varied needs (unfulfilled rights) and expect their CFUG to deliver – an expectation arising from CFUGs' success in managing forest use rights.

This argument is based upon the concept of communal access to forest resources as a *right* rather than a *policy*. Policies can change with political winds and revolving parliaments. A *right*, however, is more difficult to retract;

"once a benefit stream to the poor has been established as a right, it is difficult to reverse, and considerably easier to defend against corruption or political capture" (Conway et al, 2002, p3). If forest access is a *right*, a right to livelihoods and improved quality of life, this provides not only the needed security to maintain that stream to resource benefits, but also creates the space to demand an ever-widening circle of rights.

Placing Community Forestry

> *People do not love the government. The government took the forest from them and left nothing for the people, so the people didn't love the forest.* (Jureli CFUG member and FECOFUN district chapter chairperson, 5 October 2007)

Community forestry in Nepal began, in part, as a way to correct the negative impacts of previous government nationalization of forestlands. The 1993 Forest Act and the 1995 Forest Regulations created community forest user groups and began to devolve forest control on a larger scale, although such groups were "largely promoted by international development organizations" (Nightingale, 2003, p527). The experience of community forestry in Nepal is largely deemed a success – one that has improved forests and benefited communities on multiple fronts, including governance, and social and economic justice (e.g. Shrestha, 2001; Timsina, 2003; Kanel et al, 2004). However, even CFUGs that are successful on some measures fall short on others. Studies have documented shortcomings of the CFUG experience, including reinforcing traditional elites and inequities (Timsina, 2003; Nightingale, 2005), and succumbing to political struggles both across scale (Shrestha, 2001) and within groups (Varughese, 2000).

This chapter presents a dimension of community forestry in Nepal's Tarai that has received little attention – securing tenure rights may lead to claiming and realizing a wider set of human rights. Much community forestry literature is based in Nepal's hill regions; this is not unexpected, as about 90 percent of all CFUGs are in hill and mountain districts (ISRC, 2010). The Tarai plains are home to the remaining 10 percent (ISRC, 2010), along with around half of Nepal's population, and yet receive comparatively less academic attention. That attention tends to focus on national parks and buffer zones (e.g. McLean and Straede, 2003; Nagendra et al, 2007) or governance structures, which are highly contested (e.g. Nagendra, 2002; Ojha, 2008). After eradication of malaria sparked a forest clearing and homesteading boom during the 1960s, the Tarai's highly degraded and isolated forests were transferred to CFUGs in the 1990s. While better-quality forests did exist, those remained under government control (Nagendra, 2002).

Nepal's CFUGs are charged with managing forests for communal benefit, sustaining livelihoods, and alleviating poverty. Groups have some creative freedoms in how they achieve this. However, due to lack of internal funds and/or lack of expertise, many CFUGs choose to partner with national and international organizations to "bring programs". Although this is a reality for most CFUGs, not all CFUGs' projects depend upon these external partnerships. Thus, while many (rightly) associate community forestry in Nepal with international partnerships, this is not the only method by which users' rights are expressed and enhanced. CFUGs with diversified sources of income, such as those in this study, can independently conduct needed projects and programs, and not rely solely on partnered projects to address community needs. The CFUGs discussed below are advancing human rights and creating better lives for their communities, both through their own internally funded programs and through partnered projects.

Jhapa CFUGs and FECOFUN

Jhapa, located in the eastern Tarai, is home to 28 established CFUGs with an additional 24 in the process of being formed. While all five groups in this study rely on the sale of forest products for their budgets, some also profit from rubber tapping, tourism, and interest earned on loans to users. In 2008, the 28 CFUGs had an average income of 987,017 Nepalese rupees (around US$14,000) each and managed a total of 7685ha. The CFUGs are large in size and income compared to hill counterparts. For example, the hill district of Gorkha had 361 CFUGs with an average income of 12,912 Nepalese rupees (around US$180) and managed 16,748ha total (ISRC, 2008).

These CFUGs work in areas beyond forest management and access that improve quality of life in their communities (see Table 6.1). During a community meeting in 2007 to review the new Prakriti CFUG budget, a committee member declared:

> *Our community forest group has brought us clean water. What have the politicians in Kathmandu done for us? We have built roads and a school. What have they done for us in Kathmandu? If we want change, we need to do it as a community. We are the ones working to improve our community.*

This sentiment is echoed by community members who turn to their CFUGs for a range of services, from health to education to livelihoods. Prakriti CFUG is the most limited, financially, of the groups in this study. During a group interview in 2007 with poor households, respondents claimed inequity in the CFUG's road improvement and electricity distribution; they also expected the CFUG's help in building toilets. Another unmet demand, both in Prakriti and Junkiri, is for more women-focused programs and women's income-generating opportunities. If health and livelihood are basic human rights, then users are clearly expecting a wider range of rights access from their forest management group.

Table 6.1 *Community forest user groups (CFUGs) and their projects*

*CFUG name**	*Types of projects***
Kumari	Biogas loans, women's micro-lending, animal breeding
Junkiri	Biogas loans, development of non-timber forest products, wetland restoration, medical assistance, school fee and book assistance for the poor, clean drinking water system, road improvement, goat keeping
Prakriti	Biogas loans, clean drinking water system, road improvement, electricity, pre-school, home building for elderly poor, goat keeping
Diyo	Biogas loans and grants, savings and loan program, road improvement
Jureli	Biogas loans, road improvement, electricity, job training, animal breeding, toilet building, land distribution to landless poor

Notes: * CFUG names listed are pseudonyms.
** This list reflects the programs cited by community members and CFUG committee members during interviews.

Over 11,200 CFUGs partner with the Federation of Community Forest Users, Nepal (FECOFUN), a national organization with sub-national branches whose goal is to promote and protect forest users' rights. Many organizations work in community forestry; but this study focuses on FECOFUN, given its extensive national network, and many international organizations partner with FECOFUN to implement programs. FECOFUN is also involved in a variety of social justice and rights-based activities, including political and rights awareness campaigns.

Community Forestry: Widening the Circle of Rights

> *Nothing is going to happen without community forestry; the government is going to do nothing. The community forest is doing all the development programs. The community forest is looking after the people who are getting sick. The community forest is paying for books and fees for poor village students. Biogas is very essential. Almost every family has biogas in the village. If there are people in the village who cannot afford to pay the hospital bills, we manage to take these people to the hospital and pay for their care. Though this action committee goes and a new one comes in its place, the decisions will be implemented. Even though the individuals may change, the decisions will not be changed; the programs will continue.* (Junkiri CFUG treasurer, female, 2007)

Forest rights and users' expectations

As previously mentioned, the 1993 Forest Act and the 1995 Forest Regulations provide legal basis for CFUGs and communal forest rights. Simply declaring a right does not put it into practice; people must understand how their rights operate within given structures and who to hold accountable. To this end, FECOFUN's Jhapa District chapter conducts forest rights awareness programs throughout the district, including one in Jureli CFUG in 2007.

Facilitators meet with small gendered groups and interactively discuss the history, rules, purpose, and users' rights and expectations of the CFUG to ensure that everyone is aware of forest rights and the possibilities and limitations of community forestry.

Respondents listed access to firewood, fodder, and lumber, as well as forest protection, when asked to identify the purpose of their CFUG. People also complained about timber smuggling and poor people selling firewood collected from the community forest in the market. Maintaining access to forest products, conservation, and complaints of smuggling are not surprising responses. What is interesting is that many also identified a range of community development and poverty alleviation programs among their CFUGs' responsibilities – projects such as road building, electricity, clean drinking water, job training, healthcare, and more. When these were not provided, or were perceived to be provided inequitably, it led to user frustration. By restoring what had been severely degraded forestlands into productive forests, forests that both meet users' needs and provide a communal income, communities now expect their CFUG to provide access to a broader range of rights and services that improve their quality of life. This expectation is born of the experienced success of communal resource management; CFUGs have not only become but are expected by users to be a reliable and tangible mechanism for advancing a broader range of rights.

Widening circle of rights

As outlined above, all five CFUGs are actively engaged in various projects outside the direct realm of forest access and management. The broad suite of human rights includes rights to work, healthcare, food, housing, education, and culture: the rights needed to live a life with dignity. These CFUG actions promote and advance such types of basic human rights.

A right to "livelihood" is a more appropriate framing to consider than "work" as a basic necessity for a life with dignity. CFUGs are actively involved in securing livelihoods through multiple approaches. They employ community members in lumbering and forest-based enterprises – for example, Junkiri CFUG employs dozens of people in rubber tapping and tourism. Farming in Nepal is highly dependent upon forests; ensuring access to forest resources helps to secure those livelihoods. Two CFUGs operate animal breeding centers that also support farming livelihoods. CFUGs have dabbled in goat-keeping programs where members are given loans to raise and breed goats, but have met with mixed success.

CFUGs expand alternative livelihood options through job training, such as Jureli CFUG's sewing, hair cutting, and bamboo craft-making programs, and through micro-credit lending, such as Kumari CFUG's program to loan money to groups of five women to start vegetable vending businesses. In 2007, there were 85 groups (425 women) participating and benefiting from Kumari CFUG's micro-credit and livelihood program. Women in Junkiri and Prakriti CFUGs are demanding that their CFUGs offer more job training opportunities for women.

Not all CFUGs offer alternative livelihood programs; however, members still expect their CFUG to enhance livelihood options.

CFUGs also address healthcare. Junkiri CFUG provides need-based grants for health emergencies and pregnancy. CFUGs are involved in multiple preventative measures, including clean drinking water systems, sanitation, and biogas. Clean drinking water systems with deep wells that distribute water to communal taps, such as in Prakriti CFUG and being replicated in other CFUGs, are made possible with combined support from the CFUG and government assistance (however, Prakriti CFUG's system began malfunctioning in 2009 due to a faulty pump). Jureli CFUG provides members with supplies to build toilets. In Prakriti CFUG, a meeting with poor women revealed their sanitation concerns and frustration that their CFUG could not provide assistance in building toilets. Again, members are expecting their CFUG to play an active role in obtaining a basic human right: health.

All five CFUGs promote biogas in their villages. Biogas contributes to forest protection, improves quality of life, and improves health by replacing firewood with methane for cooking. Biogas is created by mixing manure with water and releasing it into an underground digester where the methane rises and is piped to the kitchen. While the level of support varies among groups (some work in conjunction with partners, whereas others work from their own funds), by facilitating material support for biogas, CFUGs directly affect women's and forest health. Replacing firewood for cooking improves health by reducing upper respiratory diseases suffered disproportionately by women and children who inhale smoke from open fires in enclosed spaces; a benefit cited by all biogas-owning respondents. Biogas slurry is an organic fertilizer that can replace urea, thus helping to secure farming livelihoods. By promoting biogas, CFUGs conserve forests while affecting households' health and livelihoods.

While the CFUGs' larger projects address livelihood and health, they also contribute to education, housing, and culture. CFUGs have built pre-schools (Prakriti CFUG) and provide grants to poor families for school books and fees (Junkiri CFUG). They have provided homes for the elderly (Prakriti CFUG) and transferred land to the landless (Jureli CFUG). Respondents explain that CFUGs preserve Hindu culture by providing funeral pyre wood free to the poor. Through forest management, CFUGs ensure that forest products for religious purposes are available, such as *sal* leaves. CFUGs also strive to improve quality of life by working with government offices to improve roads and electricity access.

These examples demonstrate that CFUGs have moved beyond securing forest rights and work to advance basic human rights of livelihood, health, education, housing, and culture. What would be termed "human rights" in the internationalized discourse are simply needs that are identified by the community and which the CFUGs work to attain. The FECOFUN district chairperson summarized it succinctly: "CFUG gives facilities; government can't do it. CFUGs do training such as sewing and hair cutting for the poor to start businesses. They give animals. It is organized and they save the forest."

Political rights beyond the forest

The previous section covered economic, social, and cultural rights. CFUGs also work in the areas of civil and political rights. The functioning of the CFUG itself is a demonstration of political rights (although with mixed experiences across Nepal); but that is not the focus here. Rather, this section covers political awareness campaigns, both for forest rights and for national political rights, organized around CFUGs by FECOFUN.

After the Peace Agreement was signed in November 2006 between the government and Maoists, ending a ten-year civil war, FECOFUN was among the organizations that set out to inform the broader public of the agreement, the coming new democracy, and the people's new rights and expectations. FECOFUN facilitators conducted a nation-wide campaign to explain the Peace Agreement and the upcoming election process through CFUGs. They did this through printed materials (see Figures 6.1 and 6.2), and through community meetings. Interestingly, the poster depicted in Figure 6.1 was also screened onto a T-shirt for facilitators, with the left side on the front and the right side on the back. Note that the Peace Agreement includes assurance of human rights. Figure 6.2 is a page from a flip-chart book distributed through CFUGs for facili-

Figure 6.1 *Peace Agreement?*

Note: This poster was sponsored by FECOFUN: "Elements of the Peace Agreement: Inclusive democracy, arms management, rehabilitation, and assurance of human rights, The light has embraced Nepal and the Nepalese. Now there will be sustainable peace in our country."

Source: Federation of Community Forest Users Nepal (FECOFUN)

Figure 6.2 *International Covenant on Economic, Social, and Cultural Rights (1966)*

Note: This campaign was sponsored by Canadian International Development Agency, Center for International Studies and Cooperation, FECOFUN, and Worldview Nepal.

Source: People's Voice program flip-chart (p15) illustrating the International Covenant on Economic, Social, and Cultural Rights (1966)

tators to conduct standardized awareness meetings on the new political process and human rights; included are pages explaining the International Covenant on Economic, Social, and Cultural Rights (1966) and the International Covenant on Civil and Political Rights (1966). Both campaigns were national efforts that also touched CFUGs in Jhapa.

On 15 December 2007, after the constitutional assembly elections had been postponed a second time, Jhapa FECOFUN held a bicycle rally, with CFUG members from across the district participating, to demand timely elections. About two-and-a-half years later, in April 2010, FECOFUN organized a series of Forest Caravan rallies across the country, which then descended on Kathmandu for a national rally, demanding forest rights be guaranteed in the new constitution (see Figure 6.3). Dozens of buses carried forest users and supporters to the regional rally in Biratnagar, including users from all five of the CFUGs in this study.

FECOFUN's network can reach a broad section of Nepal's population; but that is not the issue here. Rather, it is the fact that a resource management group, the individual CFUGs, would include political awareness campaigns in their activities. Politicization permeates nearly all sectors of Nepalese society – forests included – at all levels, from village to central government. 2010 has been a

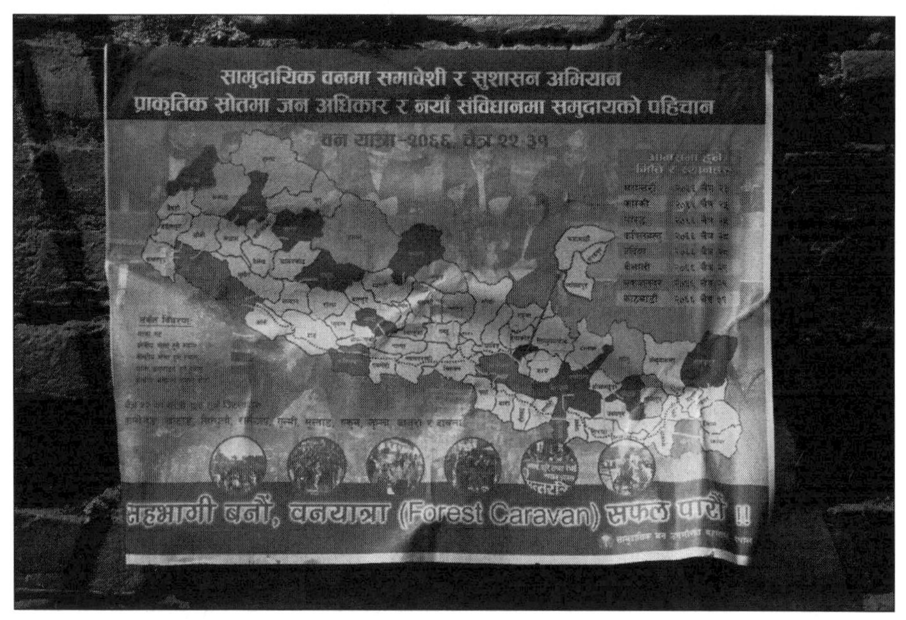

Figure 6.3 *Community Forest Campaign for Inclusion and Good Governance*

Note: This poster was sponsored by FECOFUN: "Include people's rights in natural resources and community rights in the new constitution."

Source: Federation of Community Forest Users Nepal (FECOFUN)

year of wood smuggling and forest official scandals, culminating in the government's controversial July 2010 proposal that effectively restricts forest people's rights, despite evidence that irregularities happen more often in non-community forests. Forest users view forest benefits as a communal right, not as a policy subject to shifting political winds. Echoing Conway et al's (2002) right versus policy discussion, once forest access is extended as a right, it is more difficult to restrict. CFUGs and their advocates, who have successfully campaigned for political rights, must now retrench and defend their basic right to the forest.

Connecting to human rights discourse

The preceding sections highlight the different ways in which CFUGs are promoting and enacting a broad range of human rights. Community members are making specific human rights claims, demonstrating that human rights are not a static discourse, but are modified and co-produced through action by individuals and communities according to their need. The focus is on their actions, and how those actions are examples of a locally grounded human rights agenda. CFUGs modify and enact locally understood human rights, but also reproduce larger global human rights discourses, as the next two examples demonstrate.

Over 1000 people, including school children, local organizations, FECOFUN, and CFUG members, came to the streets of Birtamod in December 2007 to celebrate International Human Rights Day, and the beginning of the 60th-year celebration of human rights. The community forest contingent carried a banner, clearly positioning themselves in the international discourse: "Let Us Successfully Celebrate the 60th International Human Rights Day Observance: Let Us Honor Human Rights." A year later, Jureli CFUG held a three-day capacity-building workshop, run by Lutheran World Federation, to train community facilitators in human rights. The trainer drew a, perhaps unfortunate, analogy: human rights must spread to all, just as mobile phones are now in every pocket. Mobile phones are only in the pockets of those with disposable incomes, in range of a network, and with the means to recharge it; mobile phones are also a foreign import, dependent on a highly centralized network. Thus, his analogy serves to demonstrate human rights criticisms – that human rights are a Western-dominated narrative in a structured network to which millions of people lack access.

When asked why the CFUG held this program, the then Jureli CFUG chairperson explained it is necessary to understand that everyone, regardless of gender, caste, and ethnicity, has the same basic rights. About 18 months later, I asked the new Jureli CFUG chairperson, elected in 2010 after a conflict arose over resource extraction that led to the previous committee being quite literally locked out by the users for eight months, about the connection between community forestry and human rights. He explained that while they do not work directly with human rights activists, "our feeling may be similar" and "we also respect the will of the people while carrying out our activities". And the will of the people is to widen the circle of human rights and for that circle to reach a wider constituency.

Conclusions

Observing these five CFUGs in Jhapa over the course of three years, it is clear that each has unique dynamics, problems, and successes. The political situation in Nepal is fluid and dynamic, and forests and forest peoples are clearly caught up in those forces. This can result in conflict and inaction (as happened in Jureli CFUG for eight months), and in focused action addressing users' varied needs – which all five CFUGs demonstrated in their various programs to improve quality of life in their communities. Some initiatives are funded internally; others are done in partnership with other organizations or government offices; but all are grounded in community members' articulations for broader services and rights. Human rights cannot be reduced to a static, universalizing, Western-bureaucratic, top-down narrative dependent on quantitative indicators. As this case shows, human rights are identified, modified, and broadened by communities. Human rights need not always be granted from on high; they can be enacted from below. The actions of CFUGs, as outlined above, demonstrate that they are actively widening the circle of rights their communities enjoy – from access to health to education to livelihoods – based upon needs identified in the community, not necessarily international agendas. Groups that find their legitimacy on a rights-based approach to forests can, and do, create their own spaces to advocate a broader social justice that pursues guaranteeing access to a range of material and social human rights, allowing a wider public to build a life with dignity.

References

Aaronson, S. A. and J. M. Zimmerman (2006) "Fair trade? How Oxfam presented a systemic approach to poverty, development, human rights, and trade", *Human Rights Quarterly*, vol 28, November, pp998–1030

Arzabe, P. H. M. (2001) "Human rights: A new paradigm", in W. Van Genugten and C. Perez-Bustillo (eds) *The Poverty of Rights: Human Rights and the Eradication of Poverty*, Zed Books, New York, NY, pp29–39

Conway, T., C. Moser, A. Norton, and J. Farrington (2002) "Rights and livelihoods approaches: Exploring policy dimensions", *ODI: Natural Resource Perspectives*, vol 78, May, pp1–6

ISRC (Intensive Study and Research Centre) (2008) *Village Development Committee Profile of Nepal: A Socio-Economic Development Database of Nepal*, ISRC, Kathmandu

ISRC (2010) *District and VDC Profile of Nepal 2010: A Socio-Economic Database of Nepal*, ISRC, Kathmandu

Johnson, C. and T. Forsyth (2002) "In the eyes of the state: Negotiating a 'rights-based approach' to forest conservation in Thailand", *World Development*, vol 30, no 9, pp1591–1605

Kanel, K. R., P. Mathema, B. R. Kandel, D. R. Niraula, A. R. Sharma, and M. Gautam (eds) (2004) *Twenty-Five Years of Community Forestry: Proceedings of the Fourth National Workshop on Community Forestry*, Department of Forest Community Forest Division, Kathmandu

McLean, J. and S. Straede (2003) "Conservation, relocation, and the paradigms of park and people management – a case study of Padampur villages and the Royal Chitwan National Park, Nepal", *Society and Natural Resources*, vol 16, pp509–526

Mertus, J. A. (2009) *Human Rights Matters: Local Politics and National Human Rights Institutions*, Stanford University Press, Stanford, CA

Mohan, G. and J. Holland (2001) "Human rights and development in Africa: Moral intrusion or empowering opportunity?", *Review of African Political Economy*, vol 28, June, pp177–196

Nagendra, H. (2002) "Tenure and forest conditions: Community forestry in the Nepal Terai", *Environmental Conservation*, vol 29, no 4, pp530–539

Nagendra, H., S. Pareeth, B. Sharma, C. M. Schweik, and K. R. Adhikari (2007) "Forest fragmentation and regrowth in an institutional mosaic of community, government and private ownership in Nepal", *Landscape Ecology*, vol 23, pp41–54

Nightingale, A. (2003) "Nature, society and development: Social, cultural and ecological change in Nepal", *Geoforum*, vol 34, pp525–540

Nightingale, A. (2005) "The experts taught us all we know: Professionalisation and knowledge in Nepalese community forestry", *Antipode*, vol 37, no 3, pp581–604

OHCHR (Office of the High Commissioner for Human Rights) (1991) *Fact Sheet No 16 (Review 1): The Committee on Economic, Social, and Cultural Rights*, July, OHCHR, Geneva

Ojha, H. R. (2008) *Reframing Governance: Understanding Deliberative Politics in Nepal's Terai Forestry*, Adroit Publishers, New Delhi

Sen, A. (2000) *Development as Freedom*, Oxford University Press, New Delhi

Sengupta, A., A. Negi, and M. Basu (eds) (2005) *Reflections on the Right to Development*, Sage Publications, New Delhi

Shrestha, K. (2001) "The backlash: Recent policy changes undermine user control of community forests in Nepal", *Forests, Trees and People*, vol 44, April, pp62–65

Timsina, N. P. (2003) "Promoting social justice and conserving montane forest environments: A case study of Nepal's community forestry programme", *The Geographical Journal*, vol 169, no 3, pp236–242

Varughese, G. (2000) "Population and forest dynamics in the hills of Nepal: Institutional remedies by rural communities", in C. C. Gibson, M. A. McKean, and E. Ostrom (eds) *People and Forests: Communities, Institutions, Governance*, MIT Press, Cambridge, MA, pp193–226

7

Forest Devolution and Social Differentiation in Vietnam

To Xuan Phuc

Forest cover in Vietnam dropped from more than 40 percent during the 1940s to 28 percent by the end of the 1980s (Nguyen Van Dang, 2001). In response to the rapid loss, the government implemented a forest devolution policy during the 1990s, under which it transferred a large area of forestland to new landholders, many of which were local households. Land-use certificates were granted to the landholders. The land-use certificate allows the holder to keep the land for 50 years, with the possibility of expansion. Five individual rights attached to the land were granted to the holders: rights to exchange, transfer, lease, mortgage, and pass the land on to third parties. The government expected devolution to be an important vehicle for the improvement of local livelihoods for the upland poor, and an effective means to put an end to swidden cultivation, which is often seen as destructive to the forest (Nguyen Quang Tan, 2006; To, 2007).

How Does Forest Devolution Contribute to Social Differentiation in the Vietnamese Uplands?

This chapter explores the contribution of forest devolution to social differentiation in the uplands of Vietnam. Based on data collected from a three-month fieldwork period in 2004 in an upland village, Thanh Cong, located in northern Vietnam, this chapter argues that the implementation of forest devolution has contributed to social differentiation. At the local level, the exercise of land rights is heavily dependent on the household's knowledge and capital, and is strongly conditioned by larger economic processes, particularly by the expansion of commodity markets for agricultural products. As a consequence, devolution provides households which have knowledge and capital with ample opportunities to accumulate land and then to derive good benefits from it. By contrast, poorer households with limited labor power and capital were unable to make productive use of the land, having no choice but to sell it to land-hungry entrepreneurial villagers. In a context where access to government credit is difficult, and where commodity markets are prevalent, the implementation of devolution has led to social differentiation characterized by differentiated access to the land among different groups of households and the emergence of labor and debt bondages. Eventually, devolution fails to improve local livelihoods for the upland poor. Furthermore, it becomes a mechanism for entrepreneurial villagers to accumulate wealth at the expense of the poor. This chapter speaks to the issues highlighted in Chapter 1 of this volume, as it deals with the question about the kinds of rights that should be developed and the origins of the rights agenda. Before the chapter proceeds with the case, it reviews some experience in forest devolution. The body of the chapter presents the case and highlights the causes, processes, mechanisms, and indicators of differentiation in the village. In conclusion, the chapter emphasizes some implications for a rights-based forestry approach and suggests some potential alternatives for mitigating social differentiation triggered by devolution.

Forest Devolution and Its Outcomes

Recent decades have witnessed a dramatic shift of control over forests, from the state to local people (Larson and Ribot, 2007; see also Chapters 2 and 5 in this volume). In Chapter 2, Sunderlin observes that during the period of 2002 to 2008, in 25 of the 30 most forested countries, the absolute area of forestland under the control of governments decreased from 80.3 to 74.3 percent of the forest area, and there was a corresponding increase in the area of land designated for use by communities and indigenous groups, in land owned by communities and indigenous groups, and in land owned by individuals and firms. Devolution is often understood as positive responses to the massive failure of centralized government ownership and control of forest estate, international solidarity movements for indigenous and cultural survival and sovereignty, and resistance and collection action at the local level, among others (see Chapter 2).

However, studies have challenged the positive relationship between individual property rights and actual benefits derived from those rights. At the local level, villagers may not be able to comprehend the rights given to them and may thus be constrained in exercising their rights (Yeah, 2004). Often, property rights are embedded within an environment of complex social relations, policies, and social actors (Verdery, 1996, 2003; Hann, 1998, 2003). As a result, property rights are only one mechanism among many, such as the use of technology, market access, and knowledge, which allow households to benefit from resources (Ribot and Peluso, 2003).

Forest devolution policy has been implemented by the Vietnamese government since the mid 1990s. By 2004, approximately 2.9 million hectares of forested land, or around 22 percent of the country's total forestland, had been transferred to households and communities. Studies examining the impacts of forest devolution upon the local population and forest environment in the Vietnamese uplands have shown both anticipated and unanticipated outcomes. In some areas, the implementation of the policy has led to improved local livelihoods, an increase of forest cover, and the strengthening of villagers' control over forestland and forest resources (Castella et al, 2006). Devolution has motivated local people to invest in perennial crops, reflecting their gradual shift to sustainable livelihoods (Castella et al, 2006). Yet, devolution has provided ample opportunities for personal gains for local elites who have connections with political power (Sowerwine, 2004; To, 2007). Quite often, benefits derived from devolution were unevenly distributed among local households, favoring those with power, knowledge, and capital (Nguyen Quang Tan, 2006). Frequently, legal rights constituted by devolution do not comprise actual rights on the ground (Tran and Sikor, 2006). Some studies suggest that the uplands have been experiencing social differentiation that is primarily characterized by differentiated access to the land and to income derived from the land. However, devolution is not directly attributed to differentiation, but to factors such as ethnicity, blood relations, resettlement history (Scott, 2002), the introduction of hybrid seeds and new cultivation techniques in agricultural production (Henin, 2002), or household cycles (Sikor, 2001).

Devolution and Agrarian Differentiation in Thanh Cong Village

Village background

Thanh Cong is a small upland village inhabited by 37 Dao[1] households, with a total population of around 200 in 2004. The villagers derive their livelihood primarily from tea and swidden crops, planted on hill slopes, and wetland rice grown in the valley. Before they settled in the current village, the villagers had lived at high elevations, with swidden cultivation as their primary source of sustenance. At that time, the land near the current village was occupied by Tam Cuu Forest Enterprise, whose emphasis was on timber logging from the forest around the village. During the second half of the 1960s the country experienced a large-scale, government-led resettlement program under which many swidden households living at high elevations, including the Dao in Thanh Cong, were asked to move to lower elevations. Thanh Cong village was formed in 1968 as a result of such a relocation effort. In 1970, the co-operative was introduced in Thanh Cong. From then on, villagers worked within a collective form of production. Management of the forest near the village still lay with the enterprise. As the area of paddy land was too small, the co-operative continued to work swidden land, often with permission from the enterprise that managed the forestland. Swidden rice and cassava were the only crops grown on collective swidden land. As collective swidden was not productive, all households in Thanh Cong worked their own swidden fields outside the collective areas in the forest. None of the households, however, asked for permission from the enterprise. During this period, agricultural production in Thanh Cong occurred primarily on a subsistence basis.

The co-operative was dissolved in 1989 and its paddy rice was redistributed among the households. Forestland allocation was implemented in Thanh Cong in 1995, so that each household received about 8.8ha on average. The land was given to the households for 50 years, with five clearly defined rights attached to the land. The land-use certificates were granted to the households between 1996 and 1997.

Expansion of markets for tea, cassava, and softwood trees

Phu Tho Province, where Thanh Cong is located, is the third largest tea-producing area in Vietnam. The province's tea production area rapidly expanded after forest devolution, from 7200ha in 1996 to 11,773ha in 2004 (MARD, 2006). Since the second half of the 1990s, tea production and its processing industry have boomed in Thanh Son District, where Thanh Cong village lies. Production is mainly household based. However, the production output barely meets 45 percent of the processing capacity of the factories located in the district (*Bao Phu Tho*, 2010). For local authorities in Phu Tho, tea has been one of the most important crops in the fight against poverty. As a consequence, substantial financial resources from provincial government and donor agencies have been spent in an effort to boost tea production in the province. Until recently, the desire to boost the capacity of the processing sector, coupled with the stable price of raw material, has served as a strong driving force for the expansion of the production area in the province.

Cassava is the second most important commodity product influencing land-use practices in Thanh Cong and other areas of Phu Tho. The area devoted to cassava production increased substantially after forest devolution, from 275,000ha in 1996 to 400,000ha in 2004 (MARD, 2006). This made the country the 13th largest cassava producer and the 4th most important exporter in the world (Ngo The Dan, 2000). The rapid expansion of the production area is attributed to the boom of the animal feed industry, in the country and abroad – most noticeably in China. By 2007, the total cassava production area in Phu Tho Province reached 7500ha, and the province became the fourth largest cassava production area in the northern region (General Statistics Office, 2009). Similar to tea, cassava production in Phu Tho is primarily household based, with cassava planted on the household's forestland granted to households under devolution. In response to market demands, the area of cassava production in Thanh Cong increased rapidly.

Softwood trees make up the third most important commodity product that has a strong influence on local land-use practices in the country, and in Phu Tho Province, in particular. Since the second half of the 1990s, paper and pulp industries, and the large-scale export of wood chips to Taiwan and Japan, have boomed (Lang, 2003). This has led to a severe shortage of raw material to feed industries. In Phu Tho Province, the Bai Bang Paper Company is the biggest paper producer in Vietnam. The province has become one of the most important softwood-producing zones for the company. Under devolution, high market demand for softwood trees has served as a driving force for local households to clear forests in order to grow softwood to sell to the company. Short on raw material, Bai Bang Company had to rent large areas of forestland from local households for plantation. This process started towards the end of the 1990s.

The expansion of commodity markets for cash crops and softwoods has played an important role in shaping local land-use practices in Thanh Cong. Local households have responded to the markets by gradually converting their forestland into tea, cassava, and acacia plantations. The production of tea received substantial support from the government, such as free seedlings and technical guidance given to households. In fact, tea production started in Thanh Cong prior to devolution, when households planted tea on gardens adjacent to their homes. However, because of the difficulty in market access, the lack of processing industry in the province and of cultivation techniques, many households had to abandon their plantations. Tea production in the village only started to increase again after devolution as a response to increased demand from the processing industry. During 1998 to 1999, about 50 percent of the households in the village had tea plantations, with about 700 to 1000 square meters each, on average. By 2002, 80 percent of the households in the village grew tea, and each of them had about 2100 square meters, on average. In 2004, almost all households in the village had a tea garden, with each of them having around 3200 square meters. Income derived from tea became the most important income source for households.

Rapid expansion of the animal feed industry in the country has substantially increased market demands on cassava. In Thanh Cong, cassava production has quickly shifted from a subsistence basis to commercial purposes. By 2004, when this study was conducted, all households in Thanh Cong engaged in cassava production for sale, with each household working 1ha to 2ha of cassava, on average, up from 0.2ha to 0.3ha during the 1970s and 1980s. Income derived from cassava sale was the second largest income source for households, as indicated by the household survey.

While the government assisted in the production of tea, the production of softwood trees is entirely shaped by market demand. Prior to devolution, none of the households in Thanh Cong planted softwood trees. This situation dramatically changed after the Bai Bang Company experienced a severe shortage of raw material, leading to the company's scramble for land in many areas in the province at the end of the 1990s. In 1999, Bai Bang Company arrived in Thanh Cong and rented land from the villagers. It was from this point on that softwood tree plantations expanded in the village. By 2008, about 60 to 70 percent of the forestland areas allocated to the households were covered by trees, mostly acacia. On average, each household had about 4.6ha of acacia. Usually, acacia is intercropped with cassava for two to three years, until the acacia closes its canopy, thus inhibiting the growth of cassava.

Devolution, combined with the expansion of commodity markets for tea, cassava, and softwood trees, has dramatically changed the ways in which villagers work the land, and the socio-economic relations surrounding these practices. This change has produced agrarian differentiation in the village, as the next section shows.

Social differentiation in the village

Household story

In 2001, Thuy and Tinh, a village couple, lived in a shelter on the edge of the village. The shelter was surrounded by a fish pond and home garden covered by tea plantation, and *Son*, a sort of resin tree, and bushes. At the time, the couple, who had two young children, did not have any valuable assets – only 3600 square meters of tea and approximately 2ha to 3ha of land planted with acacia. They did not have enough time to work their 1000 square meters of paddy land, and thus leased it to another household. Three years later, in 2004, they had about 1ha of tea and 16ha of land with acacia. In 2008, they acquired another 10ha of land with acacia. In addition, the couple shared 13ha of land planted with acacia with two households under a shared-cropping arrangement. The couple also had another 1200 young acacia trees that they had bought from another household.

The couple represents a typical example of an entrepreneurial household who benefited substantially from individual rights to the land established by devolution. When land allocation was implemented in the village, Thuy, at that time only 18 years' old and still single, did not want the land. But at the insistence of his elder brother, Thuy received a plot of 3.5ha, about half the average area each household in the village had obtained. In 1997 Thuy left the village to work for a trader in a town, 40km from the village. He helped the trader to procure cassava, bamboo, and vegetables. While working there, he met with Tinh, a Kinh[2] woman from the lowlands, who later became his wife. Thuy and Tinh left their jobs and went to Thanh Cong in 1999. They worked their 3.5ha plot and used all of their savings to buy a plot of land from a household in the village. This plot comprised 360 square meters of paddy land, 1000 square meters of fish pond, and another 1000 square meters of garden. In the beginning, they raised fish, cleared bushes, and planted tea, resin trees, and ginger in their garden.

The implementation of devolution has served as a foundation for the establishment of land markets in upland areas, including Thanh Cong village. Households with land were entitled to "exchange" or "transfer" their land to others. The land market emerged in Thanh Cong in 1999 when the Bai Bang Company came to rent 15ha of land from three village households for acacia plantation. Starting in 2002, Thuy and Tinh followed suit, using their income derived from tea and fish sales to rent and buy land from other villagers. By 2008, they were renting 16ha from six households and had bought 10ha from another two households. According to the transaction agreement, the couple could keep the land under rent for ten years, which equals one cycle of the acacia tree. The rent was fixed at 200,000 Vietnamese dong, or approximately US$15 per hectare at the time, for a period of ten years. The land-use certificate for the plot was still under the name of the owner who was entitled to the land for the remaining period of time. As a result, the couple preferred buying the land to renting it: despite the high land price, they would have much more control over the land under the former arrangement (at 3 million Vietnamese dong per hectare).

Regarding 13ha of land under a shared-cropping arrangement, the couple did not have ownership of the land. They had to cover the cost of seedlings (around US$20, or 38 percent of the total investment for 1ha of acacia), while all other costs (such as planting and weeding) were divided equally between the couple and the household who had the land. The harvest was to be divided evenly between the two sides.

Although they had obtained very little land under devolution, the couple now owns the largest land area in the village, and though not yet harvested, their trees guarantee them a huge profit in the near future.[3] This has served as a strong driving force for the couple to acquire more land; as Thuy noted: "Our business is going well nowadays. We invest all of our money in the acacia plantation. In my experience, livestock production is labor intensive and financially risky owing to uncontrollable diseases ... We are now focusing on tree planting."

Social differentiation

Devolution has provided Thuy and Tinh with ample opportunities to accumulate land. These opportunities were enabled by individual land rights and those conditioned by larger socio-economic processes at the national level. This raises the question of why the couple was able to accumulate land while others were not. There are two important factors contributing to the answer: the couple's entrepreneurial skills and availability of capital for investment. The couple acquired entrepreneurial skills while working for traders outside of the village during the 1990s. It is there that they learned about different activities associated with the trading business, particularly those related to the trading of agricultural products. They also received training in economics and gained knowledge by participating in trading networks, some of which are of use for their current business. Before working for a trader, Tinh lived in a densely populated lowland village. Her family had had to struggle to derive a living from their limited land area. Tinh did not experience land shortage when coming to Thanh Cong, an

Table 7.1 *Characteristics of the households selling/leasing out land*

Household	*Head of household's age (years)*	*School attended (years)*	*Total land received (ha)*	*Leased/ sold area (ha)*	*Number of children*	*Acacia planted (ha)*	*Area of tea plantation (ha)*	*Paddy land (m^2)*
1	26	4	5.5	2.1	2	1	0.11	648
2	31	1	9.7	6.6	3	1	0.07	1524
3	30	5	5.3	1.5	3	1.5	0.14	914
4	36	4	5.7	1.2	2	2	0.07	792
5	32	1	7.9	2.1	3	2.3	0.18	1288
Average 1	31	3	6.8	2.7	2.6	1.56	0.11	1033
Average 2	35.7	4.6	8.6	–	2	4.6	0.3	1866.4

Notes: Average 1 = average figures of the five households.
Average 2 = average figures of the 20 households under the survey.

Source: 2004 Household Survey

area she describes as "vast land and thinly populated". With a strong desire to accumulate wealth, she and her husband, Thuy, tried to put their knowledge and skills into production. This allowed the couple to acquire land from other villagers. Income derived from cassava intercropped with acacia is immediate, providing the couple with an important source of cash income to buy land and invest in acacia plantation. According to the couple's calculations, the total costs for renting 1ha of land, and for planting acacia intercropped with cassava on the land, was US$77. Harvest from cassava at the end of the year provided them with US$370. The total investment for the planting of acacia and cassava, as revealed by the couple, is not substantial; yet, why were some households in Thanh Cong unable to do this? Data from Table 7.1 helps to answer this question.

The households who leased or sold the land to the couple were young; according to the village's criteria, they were also poor,[4] they owned no valuable assets, and they had a larger number of children and smaller land areas compared to other households in the village. They did not sell or lease all of their land to the couple, but retained some of the land for themselves. They leased or sold the land because they experienced financial difficulties and needed an immediate source of income to overcome them. Although government credit sources are available, there are a number of constraints that entirely exclude the poor from accessing this loan.

Since the arrival of the Bai Bang Company in Thanh Cong, land has rapidly gained in market value and can now be sold for immediate income. For some of the poor in Thanh Cong, selling land to the couple was the only way in which they could derive a quick cash income, particularly in cases of emergency.

The land transaction between the couple and the households goes beyond a mere market transaction. Where access to government credit was heavily constrained, particularly for the poor, the couple (with their stable income from tea and other sources) was able to establish itself as a money-lender in the village. Poor households who could not access government loans and needed an immediate income had no choice but to seek out a loan from the couple. A shortage of income sources made these households unable to repay their debt in cash, thereby forcing them to repay it in land and labor. Before they sold/leased the land to the couple, all five households under the survey had borrowed money from the couple. Their inability to repay the debt in cash forced them instead to sell their land and labor. The villagers generally turn to the couple for financial assistance during times of hardship. Patron–client relationships, embedded in debt and labor bondage between two sides, have consisted of the couple being the patron and using their financial resources to provide "protection" for the households of lower socio-economic status (the client) in exchange for land and labor. All of the households who sold land to the couple were hired as laborers to work on their own piece of land. This reflects a substantial shift of the role of the former on the land.

Not only did the households who sold land to the couple enter debt and labor bondages with the couple, but so too did other poor households in the village. Again, in the area where access to formal credit was very limited, poor villagers had no choice but to seek help from the couple when a quick cash income was needed. The couple's continuing land expansion over time entails the need for a large amount of labor to work the land. Providing a loan, sometimes with low interest, to the households who needed a quick cash income has become a means for the couple to mobilize cheap labor from the villagers. Various labor arrangements have been established between the couple and the villagers. Some villagers are regular laborers, working for the couple on a full-year basis; others are seasonal workers working for the couple mainly during the planting and harvesting seasons; others work only on a day-to-day basis. The debt and labor bondages between the couple and poor households in the village have been reinforced by the couple's livelihood diversification during recent years. During 2005 to 2006, the couple opened a shop in the village to sell necessities such as rice, salt, and agricultural inputs, such as seeds and chemicals. When necessary, the couple allows the villagers to buy on credit in exchange for labor. In 2008, about one third of the households in the village were in debt to the couple, for a total amount of about 20 million Vietnamese dong (US$1200). In a place such as Thanh Cong, this is not a small amount.

The couple is an example of emerging rich farmers who, through market mechanisms and by using their skills, knowledge, and financial capital, are able to expand their land-based agricultural production and accumulate wealth. This emphasizes processes of social differentiation that have been taking place in the village. The implementation of forestland allocation has strongly deepened this process. Social differentiation in Thanh Cong is characterized not only by differentiated land access, but also by changes in the various types of relationships between the villagers and the couple regarding land.

Conclusions

This chapter has examined how forest devolution, encompassing the notion of individual rights, has contributed to social differentiation in an upland village in Vietnam. The implementation of the devolution policy in the context of rapid expansion of commodity markets for cassava, tea, and softwood trees, and constraints in access to government loans, have led to social differentiation, characterized by differentiated access to land amongst village households, and to the emergence of labor and debt bondages in the village. Poor households with capital and labor constraints are unable to make productive use of the land as envisioned by the Vietnamese government. Instead, the working of individual rights in the context of the rapid expansion of commodity markets and a lack of access to loans, has created a foundation for the emergence of a land market, which in turn provides land-hungry entrepreneurial households with ample opportunities to accumulate land and derive profits from it. Knowledgeable entrepreneurial households with access to capital are thus able to accumulate wealth generated from land formerly allocated to the poor under devolution. This scale of unexpected social differentiation, triggered by devolution, could be mitigated or avoided if the following mechanisms are adopted. First, devolution should only grant use-rights to local households, without the possibility of alienation. While this mechanism may work in the long run, with strong law enforcement at the local level, it may not work in the short run if law enforcement is weak, as local households often do not report land transactions to local authorities. Second, devolution should grant inclusive collective rights to local communities, making land accumulation by better-off households more difficult. However, while this may serve as a sound mechanism for preventing land alienation, it poses a huge constraint to individual households in accessing government credit programs if a household's land-use certificate is required as bank collateral for the loan. Third, the devolution of tenure rights should be accompanied by parallel measures, such as credit and technical supports (i.e. better access to government credit and market information regarding individual rights to the land granted to them), in order to provide complementary productive resources for the poor. The poor in Thanh Cong village would thus be able to bring their land into productive use rather than sell or lease it. Finally, devolution should separate use-rights from management rights, and grant only use-rights to individual households while giving management rights to collective entities such as a household group or local community.

Acknowledgements

This chapter greatly benefited from useful comments by the reviewer and editors of this volume. I would also like to thank the participants in the Rights-Based Agenda in International Forestry workshop held at the University of California, Berkeley, on 30 to 31 May 2009 for their comments.

Notes

1 Dao is one of the ethnic minority groups in Vietnam. Their population makes up less than 1 percent of the country's population. Most of them live in the northern upland region.
2 Kinh is the majority group in the country. They occupy about 80 percent of the country's population.
3 At the time of harvest, 1ha of acacia could generate at least US$1000 to $1500 (compared to about US$70 as a total investment for 1ha). The shortage of softwood trees as a result of the paper and pulp industry boom, as mentioned earlier, keeps wood prices stable. This guarantees a firm profit for wood producers.
4 The author conducted a participatory wealth ranking in the village in 2004.

References

Bao Phu Tho [Phu Tho Newspaper] (2010) "Phát triển cây chè trên đất Phú Thọ" ["Developing tea plantation in Phu Tho Province"], *Bao Phu Tho*, 6 September 2010, www.baophutho.org.vn/baophutho/vn/website/phong-su-ghi-chep/2012/2/12B4B8D8E37, accessed 24 November 2010

Castella, J.-C., S. Boissau, Nguyen Hai Thanh, and P, Novosad (2006) "Impact of forestland allocation on land use in a mountainous province of Vietnam", *Land Use Policy*, vol 23, no 2, pp147–160

General Statistics Office (2009) *Statistics Year Book 2008*, Hanoi, Vietnam

Hann, C. (1998) "Introduction: The embeddedness of property", in C. Hann (ed) *Property Relations: Renewing the Anthropological Tradition*, Cambridge University Press, Cambridge, pp1–47

Hann, C. (ed) (2003) *The Postsocialist Agrarian Question: Property Relations and the Rural Conditions*, LIT Verlage, Müster

Henin, B. (2002) "Agrarian change in Vietnam's northern upland region", *Journal of Contemporary Asia*, vol 32, no 1, pp3–27

Lang, C. (2003) *Pulping the Mekong*, Oxfam Mekong Initiative, TERRA, and World Rainforest Movement, http://chrislang.org/2003/10/01/pulping-the-mekong, accessed 15 May 2006

Larson, A. and J. Ribot (2007) "The poverty of forestry policy: Double standards on an uneven playing field", *Policy Sciences for Sustainable Development*, doi:10.1007/s11625-007-0030-0

MARD (Ministry of Agriculture and Rural Development) (2006) Statistics and Food Security Information, www.mard.gov.vn/fsiu, accessed 13 March 2006

Ngo The Dan (2000) "Strengthening international cooperation in cassava research and development programs", in *Proceedings of the Sixth Regional Workshop on Cassava's Potential in Asia in the 21st Century: Present Situation and Future Research and Development Needs*, Ho Chi Minh City, Vietnam, 21–25 February 2000

Nguyen Quang Tan (2006) "Forest devolution in Vietnam: Differentiation in benefits from forest among local households", *Forest Policy and Economics*, vol 8, no 4, pp409–420

Nguyen Van Dang (2001) *Lâm Nghiệp Vietnam 1945–2000: Quá trình phát triển và những bài học kinh nghiệm* [*Vietnam Forestry 1945–2000: Development Process and Lessons Learned*], Hanoi Agricultural Publishing House, Hanoi

Ribot, J. and N. Peluso (2003) "A theory of access", *Rural Sociology*, vol 68, no 2, pp153–181

Scott, S. (2002) "Changing rules of the game: Local responses to decollectivisation in Thai Nguyen, Viet Nam", *Asia Pacific Viewpoint*, vol 41, pp69–84

Sikor, T. (2001) "Agrarian differentiation in post-socialist societies: Evidence from three upland villages in north-western Vietnam", *Development and Change*, vol 32, no 5, pp923–949

Sowerwine, J. (2004) *The Political Ecology of Dao (Yao) Landscape Transformations: Territory, Gender, and Livelihood Politics in Highland Vietnam*, PhD thesis, University of California at Berkeley, Berkeley, CA

To, X. P. (2007) *Forest Property in the Vietnamese Uplands: An Ethnography of Forest Relations in Three Dao Villages*, LIT Verlage/Transaction Publishers, Berlin and London

Tran, N. T. and T. Sikor (2006) "From legal acts to actual powers: Devolution and property rights in the Central Highlands of Vietnam", *Forest Policy and Economics*, vol 8, pp397–408

Verdery, K. (1996) *What Was Socialism? And What Comes Next?*, Princeton University Press, Princeton, NJ

Verdery, K. (2003) *The Vanishing Hectare: Property and Value in Postsocialist Transylvania*, Cornell University Press, Ithaca, NY

Yeah, E. (2004) "Property relations in Tibet since decollectivization and the question of 'fuzziness'", *Conservation and Society*, vol 2, no 1, pp163–187

Part III

Whose Claims Are Considered to Constitute Rights?

In many situations, there are not only various kinds of claims at play, but also multiple social actors asserting such, as we observed in Chapter 1. The claims made by these actors may or may not be in conflict with each other. If they are, they may clash directly over particular forest resources, as in the case of direct competition for timber. The claims of different actors may also collide more indirectly as they relate to overlapping forest resources or functions, as in the case of collective territorial claims over forests versus individual claims to harvest timber. Finally, actors may assert competing justifications for their claims on forests. Forests may thus be subject to material and symbolic struggles involving various differently positioned social actors.

The following three chapters respond to the question of whose claims are considered to constitute rights. They consider a wide variety of social actors making claims on forests. These include locally present claimants, such as long-time residents, recent settlers, men and women. They also include actors located far away from the forest at stake, a situation characteristic of international efforts to preserve forest biodiversity. In addition, new actors may arise over time, demanding rights to forests in competition with established rights-holders. As a result, activists often find themselves in the situation that they are forced to make hard choices about whose claims they consider to constitute rights and to deserve support.

In Chapter 8, Moeliono and Limberg look at the competing claims of long-time residents, migrants, and park officials in East Kalimantan's Kutai

National Park. As residents, migrants, and officials assert different legitimacies to back up their claims, the conflict involves material struggles over forest resources as well as symbolic contestations about whose justifications are considered legitimate. Moeliono and Limberg suggest that a rights-based approach may offer opportunities for conflict resolution, but also point out critical stumbling blocks on the way. Park officials, residents, and migrants would need to define and recognize each other's demands. The officials would have to acknowledge the legitimacy of the other actors' claims on park resources – in particular, the critical significance of park resources for local livelihoods. Residents and migrants, in turn, would need to respect the officials' interest in conserving the environmental integrity and biodiversity of the park, pursuing their livelihoods strategies within environmental and biodiversity safeguards.

Just as in the preceding chapter, in Chapter 9, Edwards examines conflicts over forests, competition for rights, and negotiations about legitimacy, this time in England's New Forest. Similar to Moeliono and Limberg, she argues that conflicts over forests can only be resolved if one acknowledges the relative legitimacy of forest claims made by various groups, and if the resulting assignment of rights amongst competing claims is considered fair by all involved actors. Open, transparent debate is needed on how claims are considered to constitute rights, and whether such decisions should be based on human rights, primary claims, future environmental sustainability, or popularity. In addition, Edwards puts the spotlight on the influence of power differences on actors' abilities to get claims recognized as rights, as forest communities become increasingly differentiated. Conflict resolution requires deliberative forums which recognize cultural differences between social actors and do not limit participation through inherent cultural and social biases. Without an open debate, weaker claimants are likely to be taken advantage of by better-positioned actors, a finding that represents a "salutary warning" to forest rights activists around the world.

Similar to Edwards's "salutary warning", in Chapter 10, Middleton highlights how policy recognizing forest people's rights might inadvertently favor some claimants over others. Middleton argues that the gathering policy for non-timber forest products, proposed by the US Forest Service, aggravates existing inequalities in forest rights, favoring federally recognized tribes to the detriment of unrecognized California tribes. Middleton explores the ways in which tribal members, native advocacy groups, and associated allies have responded to this gathering policy, reframing indigenous stewardship and gathering rights as rights pertaining to US Forest Service lands. Middleton's account highlights the need to formulate flexible operational approaches to rights claims, accommodating the different historical and political bases for forest peoples' claims.

The three chapters, therefore, highlight the critical challenge faced by forest rights activists: whose claims should they consider to constitute rights? The challenge becomes clear as soon as one moves beyond abstract calls for forest people's rights – or the rights of "indigenous peoples and members of local communities", to quote just one of the formulaic expressions commonly used in legal texts and political campaigns. None of the three analyses offers a simple universal solution. Any efforts at determining whose claims are legitimate and deserve support require context-specific approaches which acknowledge competing material claims and contested justifications for these claims.

This insight sheds new light on a longstanding debate among forest rights activists about indigenous peoples' collective claims and indigeneity as a justification for claims on forests. Advocacy for indigenous peoples' collective claims has afforded some forest people additional leverage at the local, national, and international levels, as illustrated by Chapters 2 and 15. The emphasis on indigeneity has offered important support for people whose identities have put them in a weak position *vis-à-vis* governments, dominant cultures, and mainstream societies. Nevertheless, the privilege accorded to indigenous peoples and indigeneity may also have come at a cost to other sorts of forest people and justifications for forest claims. Some actors have proven more capable of asserting indigenous identities than others, as shown by Middleton. Assertion of indigeneity may also have afforded some actors advantages over other actors, even where the latter possess other convincing justifications for their claims, as indicated by Moeliono and Limberg. More generally, the emphasis on collective indigenous identities may have weakened the claims of other sorts of actors, such as "non-indigenous" forest people (e.g. migrants) and people asserting other kinds of individual and collective identities within or outside indigenous groups (e.g. women or poor people).

In addition, the chapters highlight that considerations of claims and definitions of rights typically involve inclusions *and* exclusions, material and symbolic. If multiple actors compete over forests, forest rights activists often create new boundaries of inclusion and exclusion by lending support to some claims and, consciously or unconsciously, denying the legitimacy of competing claims. In this way, forest rights activists may inadvertently exacerbate existing conflicts or create new ones, despite their good intentions. Yet, consideration of some actors' claims as rights does not have to deny the same status to the claims asserted by others. Similarly, support for some justifications as legitimate does not have to reject the legitimacy of other claims. The problem is that exclusive notions of rights remain deeply entrenched. Yet, the definition of rights does not have to be exclusive.

Use of inclusive rights definitions can help forest rights activists to acknowledge competing claims made by multiple actors and to consider them all as rights. As soon as one employs inclusive notions of rights, it becomes clear that considering the claims of a particular actor as rightful does not necessarily imply that the claims made by other actors have to be seen as less rightful. The mediation approaches and use of environmental safeguards discussed in the three chapters show practical ways to operationalize inclusive rights definitions. Inclusive rights definitions, therefore, may open up new possibilities for accommodating conflicting and overlapping claims on forests in their material and symbolic dimensions. Reference to inclusive rights allows rights activists to acknowledge the material claims and symbolic justifications asserted by multiple actors. Inclusive rights may also facilitate more flexible definitions and assignments of forest rights that can be adjusted over time.

8

The Challenges of Developing a Rights-Based Approach to Conservation in Indonesia

Moira Moeliono and Godwin Limberg

In 1998, the Indonesian parliament mandated the government to implement agrarian reforms and to improve natural resource management (Parliamentary Decree IX, 1998). With this decision, land and resource rights became part of the national agenda. Progress has been slow, however, especially with regard to forestlands. Although the Indonesian Constitution recognizes indigenous rights, and Decree IX mandates the government to redistribute land, implementation has been difficult and slow. The Ministry of Forestry issued several regulations recognizing different levels of rights.[1] Not only local communities demand rights to forestland. Other actors, such as private business, often compete with local communities over rights to land and resources. Moreover, with decentralization, local autonomous governments demand control over forestlands in their area. As a result, multiple claims to the same resource have muddled the issue of rights, while rights themselves have become fuzzy due to the mobility of local people.

This situation is most obvious in the 534 protected areas in Indonesia, covering a total of 28.2 million hectares, especially in the large areas of the 50 national parks. Because of their protected status, these areas contain (relatively) unexploited natural resources which are coveted by various parties. Local government views the protected areas as unrealized potential for development; people (local as well as migrants) consider the land they contain as free for the taking; and the mining industry seeks to make a profit in the name of development. On top of this, local government resents the fact that large areas within their territories are beyond their control.

Most protected areas are paper parks created as part of the overall forest estate by a national government far removed from local communities inhabiting these areas. They tend to be established by decree, often without official gazettement or operational management, and typically without consultation or knowledge of local people (Kaimowitz et al, 2003; Contreras and Fay, 2005). Nor are claims to these areas made by other actors, such as private enterprises, recognized. The national government retains authority over conservation and protected areas, but does not have the capacity to monitor, while local governments have no interest to protect these areas. As a result, most national parks today have large settlements or infrastructure developments within their boundaries. Some contain mining activities, and most suffer from illegal logging and/or collection of forest products.

One such park is Kutai National Park (KNP) in East Kalimantan. Unlike many others, there was no large settlement of indigenous people in this area when it was initially declared a wildlife reserve. In 1995, when it was designated a national park, a thriving population of migrants, mostly from South Sulawesi, was living in the area. When a new district was established in 1999, the local government proposed to excise 23,000ha from the park to accommodate these people. Although rejected by the central government, local government has since treated the area as already excised, allowing the construction of a road, a bus terminal, gas stations, and several communication towers. In addition, indigenous Dayak and Kutai people claimed and cleared an additional 1000ha in 2007, further complicating the situation.

This chapter discusses how a rights-based approach can serve the recognition of local people's rights and ensure environmental protection in such a situation. It interrogates efforts of an alliance of a research institution, international and local non-governmental organizations (NGOs), local and central government, and the park agency in Kutai National Park through a set of related questions. How can conservation be executed? How can the rights of local people be recognized? What kinds of rights are to be recognized when indigenous peoples compete with migrants? And how can local government carry out its duty to serve its people in the park without possessing any powers over the protected areas? The chapter speaks to one of the key questions identified in Chapter 1 of this book: whose claims on the forest are recognized as rights? It concludes that effective rights recognition requires a process of negotiation and dispute resolution on the basis of effective and accountable governance. Negotiation and dispute resolution lead to rights constellations in which rights are not absolute, and where the involved parties need to compromise if any rights are to be recognized in a secure manner.

The chapter is structured as follows: after a general discussion on the issue of conservation and local rights in the context of decentralization, the case of

Kutai will be introduced. The subsequent section will discuss how a rights-based approach in the form of a special use zone can help to reconcile the recognition of local people's rights with conservation in this and other parks in Indonesia. Just like Chapter 12, this chapter thus seeks to point out mechanisms helping to overcome the conflict between local rights and environmental protection. This chapter looks at the potential of special use zones, whereas Chapter 12 examines the benefits of environmental safeguards.

Protected Areas and Local Rights

Over the past three decades the perception of protected areas has evolved significantly. Indeed, the existence of protected areas itself is being questioned with three issues highlighted: whether protected areas are effective in conserving biodiversity (Bruner et al, 2001; Hayes and Ostrom, 2005; Hayes, 2006); the fact that many protected species are found outside the protected areas (Ancrenaz et al, 2007; Kaimowitz and Sheil, 2007; Sayer et al, 2007); and the issue of rights (Alcorn, 2008; Colchester, 2008; Fisher and Oviedo, 2008; Campese et al, 2009). A new understanding is emerging where protected areas are being recognized as part of the overall landscape to be managed for scientific, socio-economic, and cultural objectives (Borrini-Feyerabend et al, 2004; Hayes and Ostrom, 2005). Management includes collaboration with local stakeholders (Berkes, 2007) in order to avoid the human-rights violations against local communities generated by conventional conservation approaches (Alcorn, 2008). This has not meant that the idea of conserving biodiversity by means of protecting certain areas has been abandoned. Indeed, there are voices proposing the expansion of these areas and more stringent conservation efforts (Bruner et al, 2001; Curran et al, 2004), where protected areas are seen as the only hope for the imperiled biota of the mega-biodiverse region of the Malay Archipelago (Sodhi et al, 2008).

In response to these discussions, Indonesia has started to accommodate demands by local people whose territories have been declared protected areas. Two recent official decrees provide guidance on recognizing the roles of local communities: a 2004 decree on collaborative management (MoF, 2004) and a 2006 decree on zoning of national parks (MoF, 2006). While collaborative management allows local communities to get involved in management, decision-making remains with the Ministry of Forestry (MoF), and activities as well as use are legally quite restricted. Collaborative management is framed more as a responsibility than as a right. The decree on zoning, on the other hand, provides limited recognition of rights. In addition to the core zone and wilderness zones, which are to be strictly protected, the decree allows four zones for different levels of use by local communities (use zone, traditional zone, religious zone, and special zone), thereby recognizing the rights of communities to live within the protected area and fulfill their basic/subsistence needs.

The zoning decree may be considered to incorporate a rights-based approach to conservation, if one follows Colchester's (2008, p3) definition that such an approach involves recognition of "the rights of forest peoples to own, control, use and peacefully enjoy their lands, territories and other resources and to be secure in their means to subsistence". On the basis of this definition, we judge the zoning decree to employ a rights-based approach (RBA) where forest peoples have their basic rights and freedom to claim lands and forests in accordance with their customary norms.

In Indonesia, RBA is generally seen as strongly linked to the "*adat*" (customary law and customs) movement. In the context of national parks, some officials believe only *adat* communities should have their rights recognized and be allowed to live within park boundaries. Legal recognition of *adat* rights, however, has long been hampered by the fact that *adat* communities have to provide scientific evidence of being a jural entity (*rechtsgemeenschap*), having a clearly defined territory and functioning leadership (Appell, 1991). Few communities are so proven to exist. Yet, in practice, *adat* communities are informally recognized by local governments without any systematic evidence, and so are their rights in a limited sense. At the same time, it remains unclear what rights accrue to local communities not considered *adat* (indigenous), as Indonesia's forests and protected areas harbor a large variety of local communities and other kinds of actors.

Another question is what rights *adat* communities have. In colonial times, van Vollenhoven conceptualized the rights of *adat* communities as:

> *... the fundamental right of a jural community freely to avail itself of and administer all land, water and other resources within its territorial province for the benefit of its members, and to the exclusion of outsiders, except those to whom it has extended certain limited, and essentially temporary, privileges.* (cited in Holleman, 1981, p43)

This concept is often applied in the explanation of '*hak ulayat*', or customary rights. More recently, it is adopted by local governments trying to sort out the customary rights in their districts (e.g. two regulations by two districts in East Kalimantan: Nunukan and Malinau).

Interestingly, while accepting van Vollenhoven's concept perceiving *adat* rights as use rights, communal and inalienable (Sonius, 1981, pXLVII; Bakker, 2008), the national land law only recognizes individual property as opposed to state property, and has no patience for the intricacies of communal property (Wollenberg et al, 2006). Districts tend to recognize *adat* claims as communal but, as the case of Kutai shows, will support individual property when it suits their needs.

The Case of Kutai National Park

Kutai National Park (KNP) in East Kalimantan (198,629ha) comprises six ecological formations: ironwood–dipterocarps–kapur (*Dryobalanops*); mixed dipterocarps; heath forest; swamp forest; flooded forest; and mangroves (Wirawan, 1985). Its extraordinarily high biodiversity includes a large proportion of endemic species – for example, 8 of the 13 genus of the Dipterocarp family and half of all Bornean mammals (Wells et al, 1999), including 11 of the 13 primate species, such as orangutan (*Pongo pygmaeus morio*) and proboscis monkey (*Nasalis larvatus*), and 80 percent (330) of Borneo's bird species (BTNK, 2005).

As one of the last remaining areas of tropical lowland rainforest in East Kalimantan, its value is primarily that of a gene pool and seed bank. In addition, the area provides some important environmental services, especially as the source of domestic and industrial water for the towns of Bontang and Sangatta (Wells et al, 1999), and provides a natural buffer against industrial pollution (Sudiyono, 2005).

The park started as 306,000ha of wildlife reserve approved by the Sultan of Kutai in 1932 ("Zelfbestuursbesluit" no 80-22, dated 10 July 1936, approved by the governor of the Banjarmasin Residency on 25 July 1936). In 1957, after independence, the area was confirmed as a wildlife reserve (BIKAL, 2002). In 1982, during the third World Park Congress in Bali, it was proposed as a national park. In 1995, it became the first national park in Indonesia. By then the area had been reduced to its present area due to several developments in the region.

In 1967, three large timber concessions operated in the park. Camps and roads were built and people arrived in droves. A log pond was built within park boundaries. The state oil company PERTAMINA rediscovered Dutch oil wells and built drilling rigs. In 1974, a gas plant (PT Badak) and a fertilizer plant (PT PUPUK Kaltim) were built in the town of Bontang, with the consequent increasing population. At the same time, political unrest in south Sulawesi drove people to migrate, and some settled along the coast. The village of Sangatta, on the northern boundary of the park, rapidly grew after large-scale coal mining commenced. In 1991, a road was built to connect Bontang and Sangatta, opening access to the park. More and more people arrived and settled along this road. In 1999, Bontang became a township, while Sangatta became the capital of the new district of Kutai Timur.

Thus, in 1995, when the area officially became a national park, it was sandwiched between two growing towns and surrounded by several timber and mining concessions with an oil company operating within its boundaries and six settlements mostly populated by Bugis migrants (Nanang et al, 2004; Arnscheidt, 2009). Nevertheless, unlike many other national parks and protected areas in Indonesia, KNP was designated in an area with no permanent settlements of indigenous peoples. Dutch literature mentions only two settlements in the area, at Sangatta and Bontang (Buijs et al, 1927), both populated by

Bugis migrants from South Sulawesi, claiming to have arrived as early as 1922 (Departemen Kehutanan dan Pusat Studi Lingkungan Universitas Mulawarman, 1993). The indigenous Kutai people may have had access rights to the area; but by decree of the Sultan of Kutai, rights were relinquished to the state when it was declared a wildlife reserve (Karib Kutai, undated).

The Struggle for Local Rights

District governments see their mandate mainly to "develop" their citizens and regard conservation as contrary to development (Arnscheidt, 2009). A park within their territory is considered a hindrance, especially for the district of Kutai Timur, with its capital city right on the border with the park. A large area with potentially valuable coal and oil deposits is beyond their reach. The park itself is in a bad condition; large parts were degraded by the fires of 1982 and 1997, and there is encroachment along the road. By the end of 1997, some 15,700ha had been cleared for cultivation (Karib Kutai, undated) and three settlements were legalized as "villages" by the provincial government.[2]

In 1999, the acting district head of Kutai Timur declared his support for recognition of the (ownership) rights of the people,[3] and with support from the province requested the MoF to convert part of the park into agricultural land as an "enclave" (Nanang et al, 2004). More people arrived, including some 500 Kutai people, claiming rights given by the Sultan (*Kompas*, 2000). In 2000, the MoF approved, in principle, the excise of the area of the three villages. However, one village then demanded an additional 8012ha, which was refused by the forestry authorities. The matter was left undecided and never implemented. Over the next seven years, the issue was taken up several times, but no decision was ever reached.

In 2007, the issue of the "enclave" was revived again with the formation of a special team mandated to facilitate a solution. At the same time, a group of several hundred "indigenous" peoples arrived and cleared some 500ha almost overnight. This group mainly consisted of Dayaks from remote parts of Kutai Timur, claiming that the district government had ignored their right to develop. They were therefore forced to move to better accessible places to improve their livelihood. This crisis finally triggered action by the MoF, and a team was sent to investigate. In December 2007, a report was submitted to the minister; but as of August 2010, no decision had been made.

Meanwhile a coalition of researchers, NGOs, companies, and the national park agency, concerned about the situation, had started to discuss the problem. The special use zone concept was developed as an attempt to reconcile local rights to, and the use of resources with, the conservation function of the park.

Discussion

The coalition grappled with three issues. Which kinds of local people should receive land rights? What are the best institutional mechanisms to reconcile conflicting demands on the land? And what governance arrangements need to be in place to support negotiations among involved actors, and to resolve disputes? The following discussion takes up one issue after the other.

Who are the rights-holders?

Various local people are making claims on the parkland on the basis of different justifications. Although Buijs et al (1927) noticed no permanent settlements in KNP, Basap, and Punan Dayaks and Kutai Sangatta people probably had use rights and practiced shifting cultivation in the area at that time (BIKAL, 2002; Arnscheidt, 2009). Claims today are based on a history of settlements along the coast, some dating to 1922 or 1924 (Departemen Kehutanan dan Pusat Studi Lingkungan Universitas Mulawarman, 1993; Arnscheidt, 2009); but most are from the 1960s. Dayak and Kutai people arrived only in 2007 to 2008.

Ignoring the protected area status, the provincial government gave official status to the settlements, granting legitimacy to the settlers. This triggered continuous migration, especially during the period before the district head elections in 2001 and 2005, driven by the rumors that the candidates promised security of land (Arnscheidt, 2009). A 2008 survey showed that most (54 percent) of the 4927 households today arrived only after 2000 (Departemen Kehutanan, 2008).

The different groups interact by using different rules and visions, resulting in a cascade of claims, arguments, and potential conflicts. Even within one ethnic group rules can differ.[4] In general, customary rule provides rights based on two principles:

1 the law of *merimba*, where the first person to clear (unclaimed) primary forest obtains the rights of use in perpetuity; or
2 transfer of right from original rights-holders to new arrivals, usually in return for a sacrifice or special token.[5]

In KNP, neither rule has been followed. The Bugis cleared ‘empty’ land following the Bugis version of ‘*merimba*’, ignorant of possible indigenous claims. Indigenous Dayak and Kutai claim historical rights of access and use based on being indigenous people of Kalimantan.

Thus, the question of which local group deserves recognition of their land claims proves to be a very tricky one: is it the Bugis, who settled here at least three generations ago, or the supposedly indigenous but nevertheless recently arrived Dayak and Kutai peoples, who have not lived in the area for a long time, if ever at all? Is it the Bugis, who settled before the park was established, or the Dayak and Kutai, who cleared the forest knowing that it was a national park?

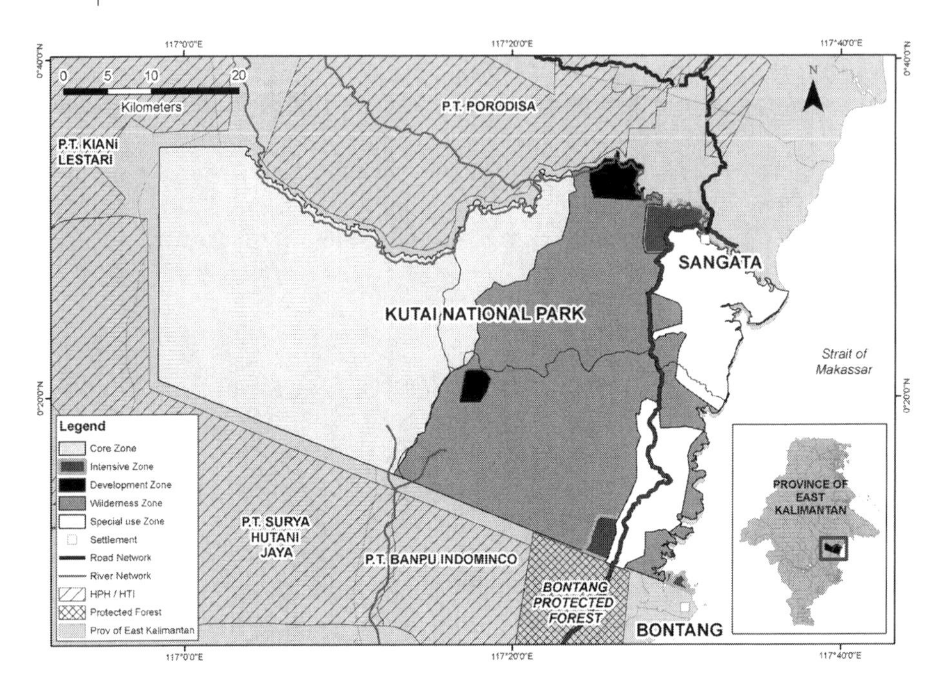

Figure 8.1 *Kutai National Park, East Kalimantan, surrounded by concessions and sandwiched between the towns of Bontang and Sangatta*

Source: CIFOR GIS, 2010

Both groups now claim "ownership" of the land, thereby also wanting to gain the right for alienation. In fact, many of them have already sold land parcels to people from nearby towns. What rights do these buyers have? The struggle over who has what rights in this protected area continues unabated.

What institution reconciles conflicting demands best?

The coalition of researchers, NGO activists, business representatives, and park officials identified three options for overcoming the conflict in KNP: evict, enclave, or establish a special use zone. Eviction of local people was ruled out quickly, as it has, in fact, never been done in Indonesia[6] and would incur horrendous financial and political costs. The enclave solution implies excising an area of the park, transferring jurisdiction over this area to the local government, and granting absolute ownership rights to the local people. This is the option favored by a large number of involved people. They appreciate the clear system of ownership rights and procedural security offered by property legislation, which, in effect, prevents any significant interference of the state (or any other outsider) on what the owner does with the land. This preference originates from indigenous and other communities' long-term experience of harsh repression over many years, and from their aversion to any responsibilities for forest protection.

The problem with the enclave solution is that there is great risk that local people will still lose the land in the name of economic development biased towards the interests of large industries and entrepreneurial actors. Often, this results in alienating land from local groups with limited compensation, while the more aggressive or better-connected migrants are able to take advantage of the situation and demand (and are given) compensation for being displaced.

The coalition has come to favor the proposal of a special use zone. A special use zone is defined as part of the park occupied by people and/or infrastructure who/which were present before the park was established (MoF, 2006) and therefore under the jurisdiction of the park agency. Rights to land will be limited to access and use rights, not ownership. On the other hand, since the people will be asked to maintain the protection function, they will have the right to receive compensation for this duty. It is envisioned that eventually local people will gain rights through the development of sustainable management systems and collaboration with the park agency. This is in line with RBAs, as the special use zone is not only about recognizing rights, but also about having the power to claim rights and about protecting and securing these rights, as well as the ability to seek redress when rights are denied (Shepherd, 2008; Blomley et al, 2009).

A special use zone within the park requires special rules where use must be compatible with conservation, and management becomes a collaborative effort. Experiences elsewhere suggest the need for development of local rules and means to enforce them, with special attention to alleviate poverty (Fisher et al, 2005; Kaimowitz and Sheil, 2007), as well as a well thought-out system of defining and recognizing rights. It is envisioned as a kind of eco-village with clearly spelled out and enforced limitations and requirements. These might include, for example, limits to population size, no ownership rights (use only), no constructions larger than a certain size, no large-scale industrial development, and a certain percentage of green area managed as commons. In return, certain facilities would be provided: healthcare, schools, electricity (using either micro-hydro or solar panels), secure management rights for individual families, support to develop sustainable farming methods, access to markets, and preferential hiring for park work.

Required governance arrangements

A special use zone requires mutual recognition of not only rights and duties, but also governance powers and accountabilities: the powers of the MoF to gazette areas of national and global biodiversity importance as public goods, beyond the direct benefits that local stakeholders obtain from environmental services (such as water for Bontang and Sangatta); the powers of the local government to provide basic service and development opportunities to all the people living within its administrative boundaries; and, of course, the rights to land and the rights to make a living for local people.

Basically, such an arrangement can only work if all parties work out a process to agree, not only on the appropriate division and recognition of rights (and acceptable limitation of rights), but also of responsibilities, including to be held liable for the consequences of their action within the area (Sonius, 1981, pXLVII; Bakker, 2008; Blomley et al, 2009).

The question is how to encourage duty-bearers to carry out their duties and be held accountable, and how to empower claims-holders to assert their rights and ensure that they understand the responsibilities/duties linked to the recognition of their claims (Blomley et al, 2009). Clearly, it should be the role of the government to ensure that duty-bearers are kept accountable, legitimate claims are honored, and all components of society, including the government, do not only claim their rights but also fulfill their responsibilities.

RBA implies that local communities have the rights and obligations to make decisions in accordance with regulations and laws. Not everyone will be comfortable to devolve this power, and not all duty-bearers will be comfortable with having this power. Regulations are only as good as the willingness of all segments of civil society to adhere to them or the capacity of the government to enforce them in such a way as to ensure that stronger sectors of society do not damage public interest (and, ideally, do not overexploit weaker actors or natural resources).

The current situation of neglect has made KNP a place of opportunities for numerous land speculators. Imposing rules and enforcing the inalienability of land would not be welcome. On the other hand, if the local government is really concerned only about the well-being of the people occupying land in the park, they should consider the special zone an acceptable solution.

Obviously, developing a special zone requires much more thought and experimentation. It requires intensive multi-stakeholder negotiations and the development of institutions to enforce agreements. It requires time, investment and consistent attention. And there is no guarantee that it will succeed. However, chances for ultimate sustainability may be better than giving it up to private property and having it eventually destroyed by mining, or maintaining the status quo because of the weak law enforcement and likelihood of further encroachment.

Conclusions

Rights-based approaches for conservation or management of natural resources work best if rights and the rights-holders are well defined and easily recognizable. They also require that parties realize two things: first, they have to recognize the rights of others if they want to achieve recognition of their own rights; and, second, different claimants have to reconcile their demands and manage an area together based on negotiated agreements. This implies for the case of national parks that park administrators need to allow rightful communities to make a living within its boundaries. These communities, in turn, need to manage the land in a manner compatible both with livelihood needs and conservation purposes.

In other words, all involved actors have to accept that rights are not absolute. Although individual interest should not be abused in the name of public good, neither should individual rights endanger the public interest. Rights always involve a mixture of private rights and obligations imposed for the benefit of public interests.

The insights from Kutai National Park also show that rights-based approaches require suitable governance arrangements, including local people's empowerment, the building of partnerships, and accountable government (Blomley et al, 2009). The exclusive power of the MoF to manage protected areas, afforded by Indonesian legislation, has no meaning if the ministry cannot effectively exercise this power. The MoF is being challenged and will have to learn how to share its powers to protect Indonesian mega-diversity and environmental needs, not only by developing the right legal framework, but by inviting other sectors of civil society to try to interpret and apply these regulations in the different social settings. Local government, in turn, needs to understand that regional autonomy is not independence. Its authority to govern a district is not absolute.

Notes

1 Options under the law include the rights to manage customary forests and village forests, as well as communal management of production forest. Recently, communities are invited to apply for leases of up to 100 years as part of the industrial plantation efforts.
2 Did the provincial government have this right or violate a law? Intuitively, one would say that no new villages should be established in a conservation area. Since nobody contested this act of the provincial government, its right was *de facto* acknowledged to provide legal certainty to the people and to provide for development.
3 To achieve this, the status of the land would have to be changed from conservation area to other use area (non-forest use area).
4 In the district of Malinau, East Kalimantan, indigenous Dayak groups dispute each other's rights based on different times of arrival in the area (Moeliono and Limberg, 2009).
5 Information from discussions with Ramses Iwan, Dayak Kenyah from Malinau, and Thomas Neet, *adat* leader from Manggarai.
6 Pers comm, Wiratno (workshop on special use zone, Bontang, 26 August 2010).

References

Alcorn, J. B. (2008) "Why forest conservation is not good news for local communities", *Arborvitae*, vol 36, p7

Ancrenaz, M., L. Dabek, and S. O'Neil (2007) "The costs of exclusion: Recognizing a role for local communities in biodiversity conservation", *PloS Biology*, vol 5, no 11

Appell, G. (1991) "Resource management regimes among the swidden agriculturalists of Borneo: Does the concept of common property adequately map indigenous systems of ownership?", Paper presented at the International Association for the Study Common Property Conference, University of Manitoba, Winnipeg, Canada, 28 September 1991

Arnscheidt, J. (2009) *Debating Nature Conservation: Policy, Law and Practice in Indonesia: A Discourse Analysis of History and Present*, Leiden University Press, Leiden

Bakker, L. (2008) *"Can We Get Hak Ulayat?": Land and Community in Pasir and Nunukan, East Kalimantan*, University of California at Berkeley, Center for Southeast Asia Studies, http://escholarship.org/uc/item/5pj3z2jr, accessed 4 March 2010

Berkes, F. (2007) "Going beyond panaceas – special feature: Community based conservation in a globalized world", *Proceedings of the National Academy of Sciences*, vol 104, no 39, pp15188–15193

BIKAL (2002) *Membangun Kepercayaan Masyarakat: Menyelamatkan Taman Nasional Kutai. Laporan Studi Sosial dan Ekonomi Masyarakat di Kawasan Taman nasional Kutai, Kalimantan Timur*, Yayasan Bikal, Samarinda

Blomley, T., P. Franks, and M. R. Maharjan (2009) *From Needs to Rights: Lessons Learned from the Application of Rights Based Approaches to Natural Resource Governance in Ghana, Uganda and Nepal*, Rights and Resources Initiative, Washington, DC

Borrini-Feyerabend, G., A. Kothari, and G. Oviedo (2004) *Indigenous and Local Communities and Protected Areas: Towards Equity and Enhanced Conservation – Guidance on Policy and Practice for Co-Managed Protected Areas and Community Conserved Areas*, Best Practice Protected Area Guidelines Series No 11, IUCN, Gland

Bruner, A. G., R. E. Gullison, R. E. Rice, and G. A. B. da Fonseca (2001) "Effectiveness of parks in protecting tropical biodiversity", *Science*, vol 291, no 5501, pp125–128

BTNK (Balai Taman Nasional Kutai) (2005) *Data Dasar Taman Nasional Kutai (Baseline Data for Kutai National Park)*, BTNK, Bontang

Buijs, D.W., H. Witkamp, and F. H. Endert (1927) *Midden-Oost Borneo expeditie 1925*, Floras of Asia

Campese, J., T. Sunderland, T. Greiber, and G. Oviedo (eds) (2009) *Rights-Based Approaches: Exploring Issues and Opportunities for Conservation*, CIFOR and IUCN, Bogor

Colchester, M. (2008) *Beyond Tenure: Rights Based Approaches to Peoples and Forests – Some Lessons from the Peoples and Forest Programme*, Rights and Resources Initiative, Washington, DC

Contreras-Hermosilla, A. and C. Fay (2005) *Strengthening Forest Management in Indonesia through Land Tenure Reform: Issues and Framework Action*, Forest Trends, Washington, DC

Curran, L. M., S. N. Trigg, A. K. McDonald, D. Astiani, Y. M. Hardiono, P. Siregar, I. Caniago, and E. Kasischk (2004) "Lowland forest loss in protected areas of Indonesian Borneo", *Science*, vol 303, no 5660, pp1000–1003

Departemen Kehutanan (2008) *Laporan Survey Team Terpadu Percepatan Penyelesaian Masalah Kutai*, Unpublished report

Departemen Kehutanan dan Pusat Studi Lingkungan Universitas Mulawarman (1993) *Laporan Penelitian Studi Sosial Antropologi Masyarakat Pemukim dan Penggarap dalam Taman Nasional Kutai*, Universitas Mulawarman, Samarinda

Fisher, B. and G. Oviedo (2008) "Rights-based approaches to forest conservation", *Arborvitae*, vol 36, p8

Fisher, R. J., S. Maginnis, W. J. Jackson, E. Barrow, and S. Jeanrenaud (2005) *Poverty and Conservation: Landscapes, People and Power*, IUCN, Gland, Switzerland, and Cambridge, UK

Hayes, T. (2006) "Parks, people, and forest protection: An institutional assessment of the effectiveness of protected areas", *World Development*, vol 34, pp2064–2075

Hayes, T. and E. Ostrom (2005) "Conserving the world's forests: Are protected areas the only way?", Paper presented to the Conference on The Law and Economics of Development, Trade, and the Environment, Indiana University School of Law, Indianapolis, 22 January 2005

Holleman, J. F. (1981) *Van Vollenhoven on Indonesian Adat Law*, Martinus Nijhof, The Hague

Kaimowitz, D. and D. Sheil (2007) "Conserving what and for whom? Why conservation should help meet basic human needs in the Tropics", *Biotropica*, vol 39, no 5, pp567–574

Kaimowitz, D., A. Faune, and R. Mendoza (2003) "Your biosphere is my backyard: The story of Bosawas in Nicaragua", *IUCN CEESP Policy Matters*, vol 12, pp6–15, www.iucn.org/themes/ceesp/publications/publications.htm

Karib K. (undated) *Potret Taman Nasional Kutai*, BIKAL, Samarinda

Kompas (2000) "Taman Nasional Kutai menuju Gerbang Kehancuran" ["Kutai National Park at the gate to destruction"], *Kompas*, 24 February

Moeliono, M. and G. Limberg (2009) "Tenure and *adat* in Malinau", in M. Moeliono et al (eds) *The Decentralization of Forest Governance: Politics, Economics and the Fight for Control of Forests in Indonesian Borneo*, Earthscan, London

MoF (Minister of Forestry) (2004) *Permenhut 19/2004 tentang Kolaborasi pengelolaan kawasan suaka alam dan kawasan pelestarian alam* [*Minister of Forestry Decree no 19/2004 on collaborative management of protected areas*], Departemen Kehutanan, Jakarta

MoF (2006) *Permenhut 56/2006 tentang Pedoman zonasi taman nasional* [*Minister of Forestry Decree no 56/2006 on zoning of national parks*], Departemen Kehutanan, Jakarta

Nanang, M., M. Arifin, and M. Setiawati (2004) "The community of Teluk Pandan and issues of forest management in Kutai National Park", *Indonesia Country Report*, pp103–116

Sayer, J., B. Campbell, L. Petheram, M. Aldrich, M. Ruiz Perez, D. Endamana, Z.-L. Dongmo, D. Nzooh, L. Defo, S. Mariki, N. Doggart, and N. Burgess (2007) "Assessing environment and development outcomes in conservation landscapes", *Biodiversity and Conservation*, vol 16, no 9, pp2677–2694

Shepherd, G. (2008) "Forest restoration, rights and power: What's going wrong in the *ngitili* forests of Shinyanga?", *Arborvitae*, vol 36, p3

Sodhi, N. S., G. Acciaioli, M. Erb, and A. K. Tan (eds) (2008) *Biodiversity and Human Livelihoods in Protected Areas: Case Studies from the Malay Archipelago*, Cambridge University Press, Cambridge

Sonius, H. W. J. (1981) "Introduction", in J. Holleman (ed) *Van Vollenhoven on Indonesian Adat Law*, Martinus Nijhof, The Hague, ppXXIX–LXVII

Sudiyono (2005) "Bab 5. Kajian Kerusakan Lingkungan Taman Nasional Kutai sebagai Dampak Aktivitas Masyarakat dan Pengembangan Wilayah dalam Era Otonomi

Daerah di Kalimantan Timur", in J. Haba (ed) *Problematika Taman Nasional di Indonesia, Kondisi Obyektif Empat Taman Nasional*, PMB-LIPI, Bogor

Wells, M., S. Guggenheim, A. Khan, W. Wardojo, and P. Jepson (1999) *Investing in Biodiversity: A Review of Indonesia's Integrated Conservation and Development Projects*, World Bank, East Asia Region, Washington, DC

Wirawan, N. (1985) *Kutai National Park Management Plan 1985–1990*, World Wildlife Fund/IUCN, Bogor

Wollenberg, E., G. Limberg, R. Iwan, R. Rahmawati, and M. Moeliono (2006) *Our Forest, Our Decision: A Survey of Principles for Local Decision-Making in Malinau*, CIFOR, Bogor

9

Rights Evolution and Contemporary Forest Activism in the New Forest, England

Victoria M. Edwards

In September 2008, in a quiet, affluent corner of southern England known as the New Forest, there appeared to be the makings of a revolution. Notices erected across the Forest, illustrating a pony, had the initials "F U" emblazoned across its rear, representing the logo of the newly formed "Forest Uprising" group. The group had been formed hurriedly in response to the release of a consultation draft of the National Park Plan for the New Forest. The plan, a statutory document to which all statutory bodies must adhere, essentially formed the manifesto of the recently constructed New Forest National Park Authority, a part-elected, part-appointed panel of members established to govern the area on its designation as a national park in 2005. To all intents and purposes, the Forest Uprising looked like a classic challenge from long-established forest users to a Johnnie-come-lately governing authority. Indeed, any outside observer might see it as a case of traditional forest activists resisting redistribution of rights or protesting their lack of participation in political decision-making procedures. However, although the Forest Uprising showed all the signs of a forest rights campaign for equity recognition and participation, appearances can be deceptive.

The New Forest National Park encompasses 93,094 acres (37,675ha) and is a mosaic of heaths, common pasture, ancient woodland, enclosed plantations, streams and river valleys (NFNPA, 2008). It is one of the most complex, multiple-use, and multiple-rights forests in existence. Half of the land is owned by the Crown and managed on its behalf by the state department of forestry (the Forestry Commission). The other half is privately owned, in the form of settlements, farms and agricultural estates. The unenclosed pasture woodlands (termed "Open Forest") on the Crown land are rare in Britain, where 90 percent of ancient forests have been destroyed.

The New Forest was named "*Nova Foresta*" by William I ("the Conqueror") in 1079 and, as such, is neither new nor exclusively a forest. At the time, the term "*foresta*" referred to an extensive tract of land, including both woodland and pasture, reserved for the Crown's privilege as a hunting ground and governed by Forest Law. The system of agriculture was "silvo-pastoral" (agro-forestry), with pasture interspersed with broadleaved woodland where animals graze and wood is gathered. Rights-holders were known as "commoners", using the forest communally for grazing and other purposes. The term "commoner" is still used today in the New Forest to describe a resident who exercises such common rights.

Throughout its history, the commoners have been viewed as a necessary but not always welcome component of the New Forest. Indeed, for over 1000 years, the monarchy and the state have manipulated the commoners' rights in order to further their own objectives for the New Forest in a manner frequently observed in other forests around the world. Latterly, such manipulation has extended to a growing "middle" class, whose interest in the New Forest is one of amenity rather than productive capacity.

This chapter, through explanation of its long-term governance, reveals the nature of the various stakeholders' claims to the New Forest over time and examines how commoners' rights in the New Forest eventually became formally recognized and protected. However, the establishment and legal recognition of those rights in the New Forest did not come about as the result of a democratic deliberative process, as proposed in Chapter 1 of this book. As such, rights in the New Forest have always been skewed in favor of a dominant group: first, the monarchy, then the nobility, and later the state. Recent activism in the New Forest suggests that rights are now being skewed in favor of local amenity-users, once again diluting the indigenous common forest rights. However, throughout history, the commoners' forests rights have played a vital role in the New Forest's continued existence. First, formal and, later, statutory recognition of commoners' rights have enabled higher powers to achieve their own objectives for the New Forest. The commoners' presence has been tolerated because their land management practices in the New Forest have provided services complementary to the dominant uses.

As the rest of the world begins to recognize forest peoples' collective rights to cultural and political self-determination, the New Forest provides an empirical study of exploitation which, to some extent, continues today. While the New Forest now incorporates formal institutions to protect forest rights and allow rights-holders access to deliberative processes, in practice such institutions continue to be hijacked by third-party activists with their own agendas. As such, the New Forest presents a salutary warning to the forest rights agenda.

As forests develop around the world, increasingly diverse forest communities present an interesting challenge for resource governance and management. First, those establishing governing institutions need to be able to recognize the relative legitimacy of claims to the forest of different groups. Second, political decisions need to be made as to the fair distribution of rights amongst competing claims. Open, transparent debate is needed on how such distributive decisions might be made and whether they should be based on human rights, primary claims, future environmental sustainability, or popularity. Without such open debate, legitimate rights-holders will continue to be exploited by articulate third-party activists. Third, decision-makers need to ensure that the deliberative forums, which play a vital role in institutional arrangements, recognize the cultural differences of groups of stakeholders and do not limit access to decision-making through cultural and social bias (for further discussion of this point, see Steins and Edwards 1999a, 1999b).

This chapter is structured in four parts. First, an historical account of the 1000 years' evolution of institutional development in the New Forest is given. This explains how the commoners' rights and continuous stewardship of the New Forest were exploited by other more dominant users throughout history. Second, an account of the contemporary governing institutions that were ostensibly established to protect the commoners' forest rights and provide more equitable access to decision-making arenas is given. It draws attention to the ability of certain stakeholders to hijack decision-making arenas. Third, the recent "revolt" referred to in this introduction is explained, showing how it represents just another manifestation of the continuous exploitation of the commoners' forest rights by third-party activists for their own agendas. The chapter concludes with an evaluation of the lessons learned from the New Forest for the international forest rights agenda.

Historical Analysis of Rights Evolution

In most of the New Forest's history, local priorities have taken second place to national objectives for the area. The forest's institutional history can be categorized into six stages. A key characteristic of each period is the formal recognition of forest rights so that they might be subsequently manipulated:

1 *1079–1184:* Naming of the New Forest and claiming of absolute title by the king. Imposition of draconian rules, dedicating the area for the king's hunting ("Forest Law").
2 *1184–1279:* Formal recognition of common forest rights, subsequent extinguishment of those rights over specific areas ("disafforestation"), and determination of a boundary.
3 *1279–1483:* Statutory extinguishment of forest rights to allow enclosure and protection of coppice woodlands for the Crown's harvesting ("encoppicement").

4. *1483–1698:* Statutory recognition of forest rights and subsequent temporary extinguishment of those rights over successive areas in order to facilitate timber production ("rolling power of inclosure").
5. *1698–1877:* Statutory establishment of governing institutions to protect forest rights ("Verderers' Court").
6. *1877–1949:* Formation of the Commoners' Defence Society and improved participation in decision-making for commoners.

This section examines each period in turn, explaining how recognition of the commoners' forest rights allowed the monarchy to restrict, dilute, and extinguish those rights over time until, eventually, third-party activists ensured protection of the rights.

1079–1184: Imposition of Forest Law by the Crown

When the New Forest was named "Nova Foresta" by William I in 1079, the king assumed absolute title of the land and introduced Forest Law. Forest Law comprised a separate legal system, with its own courts and officers, whose responsibility was to protect the "venison and vert" for the king's hunting pleasure. The first written document of Forest Law (the *Assize of Woodstock*, 1184) documents severe restrictions preventing the taking of deer: the possession of bows and arrows; the keeping of unlawful dogs; setting of traps for deer; and hunting at night (Stagg, 1984). Grazing was allowed as it kept down undergrowth; but stock had to be removed during the "Fence Month", when the deer gave birth, and during "Winter Heyning", when forage was scarce and fawns were vulnerable to disturbance and predation (Tubbs, 1986). Commoners were forbidden to enclose their land for cultivation ("*purpestures*") or to convert woodland to arable cultivation ("*assarting*"). Felling of timber was forbidden except under the supervision of a Crown forester. In this respect, although the king recognized commoners' forest rights, he ensured that they were subordinate to his own, establishing sovereign rule for the New Forest but benefiting from the land management practices that the commoners provided. The two principal rights that exist in the New Forest today reflect those originally formalized by Forest Law:

> Right of pasture: *extended to ponies, cattle, donkeys, mules, with limited rights for pigs, and, in some parts, sheep. The majority of animals turned out for grazing comprise ponies and cattle: 4834 and 2539, respectively in 2009.* (NFCDA, 2010)

> Right of mast: *the right to turn pigs out in the* pannage *season (autumn) to feed on green acorns and beech mast, which are poisonous to cattle and ponies when consumed in large quantities. 324 pigs were turned out in 2009.* (NFCDA, 2010)

Rights that have been extinguished or abandoned since the later 19th century due to increased availability of substitutes through improved road and rail networks include:

- *right of marl* (to dig clay, used to dress the acidic soil of a commoner's holding);
- *right of turbary* (to cut turf for fuel); and
- *right of estover* (to gather branch wood from felled trees).

1184–1279: Disafforestation to raise revenue for the Crown

During the 12th century, priority for the New Forest changed from one of hunting to revenue-raising. King Richard I introduced the process of "disafforestation", which referred to the removal of Forest Law from tracts of land so that it might be cultivated by barons, in return for payment to the king (Kenchington, 1944). "Assarting" (deforestation and cultivation of tracts of land) was also permitted in royal forests that were not freed entirely from Forest Law, and fines were imposed on unofficial assarting, raising further revenue. King John (1199 to 1216) was particularly ruthless in the fines he exacted from his royal forests and in other forms of revenue-raising, including the grant of around 2000 to 2500ha of land to the Cistercian Order for the founding of an abbey in the New Forest. Vast stretches of Crown land were freed from Forest Law after the signing of the *Magna Carta* (1215 to 1225) and the *Carta de Foresta* (1217). Nevertheless, relative to other royal forests, the New Forest remained remarkably intact, probably attributable to its hunting quality, but also because the land was so poor that no one would have been willing to pay highly for its cultivation. In 1279 a boundary ("perambulation") was established, placing holdings on the periphery clearly outside the New Forest and Forest Law. Within the New Forest, sovereign rule continued and commoners' forest rights were diluted by the reduction of land available to them.

1279–1483: Encoppicement to protect timber for the Crown

Although the New Forest retained its size from 1300, changes within the forest began to change its character, as the priority for the royal forests moved from hunting to timber production. The first piece of forestry legislation in the country, the *Encoppicement Act 1483*, gave Crown officials permission to enclose coppiced woodland in the royal forests. Increased demands for timber resulted in further enactments to allow large areas in the New Forest to be enclosed. The *Orders and Rules of the New Forest* established a local system for executing royal warrants (authorizations) for deer, timber and the enclosure of coppices (Tubbs, 1986). The new rules established several Crown-appointed positions in the New Forest for local governance in a very hierarchical fashion. Freeholders of the county (i.e. free landowners) were entitled to elect "*Verderers*" and "*Regarders*" to monitor and enforce boundaries, and to supervise the marking, felling, and selling of timber (Kenchington, 1944; Tubbs, 1986).

Commoners' forest rights continued to be eroded by reduction of the area to which they had access, as land barons established their status in the New Forest as the local judiciary and king's representatives.

1483–1698: Timber production by the Crown

Land continued to be enclosed by the Crown for timber protection, and trees from both enclosed and open woodlands were felled for commercial return. While the noblemen received benefits from timber production in the form of new offices, the commoners' livelihoods continued to be eroded not only by enclosure, but also by rescission of their rights. Depletion of the forest resulted in increasingly harsh timber regulations and when James I succeeded in 1603, he instigated the first form of timber cultivation in the New Forest, ordering the collection of acorns and the plowing of land for their planting. Nevertheless, a shortage of revenue soon caused him to exploit the timber resources of the forest faster than his program of timber production: a survey made in 1608 indicated 123,927 trees "fit for felling", and yet by 1632 "there was little above 2,000 serviceable trees in the whole Forest" (Wise, 1863, p45). Corruption was rife, as Crown employees were not paid and took timber and other woodland resources in lieu of wages. By the late 17th century, the New Forest was dominated by "wind-shaken" and "decayed" trees (Wise, 1863, p44; Vesey-FitzGerald, 1977, p94). A report from the *Regarders* stated that the New Forest was "in danger of being destroyed" and an Act for the Increase and Preservation of Timber in the New Forest was passed in 1698. For the first time, the act gave *statutory* recognition to commoners' rights in the New Forest, but also provided for the immediate planting of 2000 acres and for a further 200 acres per annum, allowing "inclosure" of 6000 acres at any one time. This "statutory inclosure" gave the state the right to temporarily extinguish commoners' rights from an area of land and exclude their grazing stock while it was fenced and planted for timber. The areas that were enclosed for forestry were later reopened to grazing animals once the trees were established. However, as a woodland was opened up, the state had the right to move on to a new area, equivalent in size; thus, the process became known as the "rolling power of inclosure" and marked a significant point in the New Forest's history by establishing the *statutory* power of the Crown to override the ancient common rights.

1698–1877: Establishing institutions to protect forest rights

Until the 18th century, the inhabitants of the New Forest comprised two distinct social groups: the local land lords and the commoning tenants and freeholders. During such time, commoners had neither the legal status and social standing, nor the educational training and aptitude to defend their forest rights. However, the community began to expand in the 18th and 19th centuries as small-scale enclosure, both legal and illegal, took place. At first, small parcels of land were enclosed by a peasant class moving to the New Forest to establish a smallholding and participate in other illicit trades. Mudie (1838) reports of the smugglers and plunderers occupying the New Forest at the time and of the problems encountered by the vigilant keepers in evicting such people who were "as ready to become murderers as thieves". Cobett (1830) is equally dismissive of a second type of resident in the New Forest – wealthy merchants – who bought land from the king.

Eventually, the mismanagement of rights in the New Forest and other royal forests was addressed by a public inquiry, The Royal, New and Waltham Forests Commission of 1849, which concluded that "the present state of the New Forest in this respect is little less than absolute anarchy" (Reports of Commissioners, 1850, p357, cited in Wise, 1863, p46). One recommendation was that deer should be removed from the New Forest and the Crown should be "compensated" by being permitted to inclose yet more land. The subsequent Deer Removal Act 1851 allowed the Crown to inclose a further 10,000 acres (4000ha) and, together with the New Forest Act 1854, allowed for the first Register of Commoners and Common Rights.

The era heralded the real beginning of the challenge to the Crown's priorities for the New Forest. The fight against Crown supremacy was supported by outside amenity interests, stimulated by additional interest in the New Forest by the construction of a railway line in 1847, opening it up to recreation and tourism. A New Forest Association was established in London in 1866 by owners of the larger private estates of the forest. At the same time, successful manufacturers and traders who had moved to the New Forest became new "intellectual commoners" (Kenchington, 1944, p87). They did not occupy large landed estates, but much smaller holdings, and were not dependent upon common rights for their main source of income, but enjoyed the rights, and the horsemanship and hunting that went with the commoners' lifestyle. The attraction to the New Forest by these "amenity" residents is well documented by Hutchinson (1907, pp188–203), who is clear that a new class structure was emerging. The forest had always had its share of the landed gentry; but it was now becoming the pleasure ground of the leisured *middle* classes. Those with access to the amenity value of the forest saw common rights as an important component of the forest's management regime: the commoners' stock maintained the forest in a manicured state and therefore contributed to its aesthetic value and facilitated use of the forest for recreation by rendering it more penetrable. More importantly, the commoners' rights provided a useful tool for preventing the Crown from establishing further timber production in the forest and destroying its natural appearance. In uniting their own interests with those of the commoners, the amenity users of the forest protected the commoners from their longstanding competitor, the Crown.

The new amenity stakeholders campaigned for the New Forest Act 1877, which established the basis for contemporary governance of the area. It reconstituted the ancient Court of Verderers as a body to oversee commoners' rights and the management of the common land. The Verderers were prescribed two specific functions:

1 to act as an administrative body of matters concerning the commoners; and
2 to act as a judicial body, with powers to try offences against the forest.

The Verderers were expected to design by-laws governing the exercise of common rights in the New Forest and to impose fines for violation of any by-laws, using money from the sale of land for the Southampton–Dorchester railway. The Verderers appointed four *Agisters* to assist them in the control of commonable stock and levied charges from the commoners in order to recover any ongoing costs.

1877–1949: Involvement of commoners in decision-making

Although the reconstituted Verderers' Court represented the first real protection of common rights in the New Forest, in terms of participation in decision-making, the Court reflected inequalities of the time. Verderers were elected from the educated classes: each elected Verderer was obliged to own a minimum of 75 acres of land, in or about the forest, to which the common right of pasture was attached. This rule severely restricted the choice of Verderers in the inter-war years "to no more than a handful of aging landowners" (Pasmore, 1977, p158). Between 1877 and 1949, there were only four contested elections for Verderers. It is possible that an agreement between the major landowners in the New Forest prevented election of candidates outside their circle. One unsuccessful candidate, Captain Cecil Sutton, wrote:

> *The election in which I personally was involved clearly showed that the matter was not decided by the commoners, but bus loads of chauffeurs, cooks, gardeners, parlour maids, governesses and all sorts of good folk who have not the slightest interest in our commoners' rights.* (Pasmore, 1977, p32)

Eventually the commoners began to organize themselves, but largely in response to a threat to their livestock through lack of grazing. The commoners focused on improving the resilience of the ponies' breed and the quality of grazing in the New Forest (e.g. through drainage). Through these two agrarian initiatives, they formed The Society for the Improvement of the Breed of New Forest Ponies in 1891 (later to become the New Forest Pony Breeding and Cattle Society) and the Commoners League in 1909 (later to become the Commoners Defence Association).

Development of the industrial and manufacturing base around the New Forest's boundaries further expanded settlement in and around the forest and provided employment and a secure income for the forest's existing commoners. The desirability of the commoners' holdings as residential properties and "hobby farms" to commuters and retired residents increased as the area surrounding the New Forest was developed. Many commoners continued to turn out stock, but more as a means of supplementing their income, because of their general interest in livestock and through the desire to sustain a longstanding family tradition. The New Forest Act of 1949 reconstituted the Court of Verderers to comprise ten members by increasing the number of elected Verderers to five and introducing five additional Verderers appointed by government departments. Elected Verderers now only have to occupy 1 acre of land (0.4ha) and voting rights are vested in commoners only, and not all parliamentary electors of the district, as had previously been the case. This change in the qualification of voters was stimulated by increase in urban development around the New Forest, whose residents might swamp the commoners' votes and dictate the elected representation of the Court. There was a clear intention to limit participation in the decision-making of the Court to those with commoners' rights.

Chapter 1 of this book identifies that "the transfer of tenure to land and connected resources is the key strategy to overcome people's exclusion from forests". The above historical account of the New Forest suggests, however, that such tenure must be absolute in order to ensure proper protection of forest peoples. While there is no doubt that statutory recognition of forest rights established some security for New Forest commoners, the subordinate nature of them relative to the king's absolute title meant that they were insufficient to fully protect the commoners. The rights needed to be backed up by appropriate and just institutions which were capable of defending them against erosion or dilution. However, the commoners' defense of their rights was not fully secured once such institutions were established in 1877. As the following section reveals, governing institutions and associated rights can be hijacked by third parties eager to protect their *own* interests.

Contemporary Institutions and the Wider Community of Interest

Although half of the Verderers are still elected by those with common forest rights, the Court of Verderers has become an important tool for a much wider community in continuing to protect the New Forest from unwanted development. The forest amenity groups that had been set up by the landed gentry of the New Forest were eventually taken over by the middle-class residents, the New Forest Association being the most active in influencing the evolution of the institutions for governing the commons and the rest of the New Forest. The newcomers have continued to be the most cohesive and vociferous lobby in the control of development within and around the New Forest. In 1972 this "amenity voice" defeated the Forestry Commission's intentions to plant coniferous woodland in the New Forest. Since 1974 it has prevented a succession of proposals from Hampshire County Council to construct a road bypass around Lyndhurst, which could only be built at the expense of a major incursion into the forest. In 1982 it prevented Shell UK Ltd from carrying out exploratory drilling for hydrocarbons in the New Forest. In 1988, it resisted an attempt to construct a second power station on the New Forest's eastern edge, and in 2004 it opposed the construction of a large container port nearby. In most of the above cases, the voluntary organizations (18 in the case of the Shell inquiry) came together under the umbrella of the New Forest Association to present a collective case at the respective public inquiries and meetings. Interference with commoners' forest rights was always cited as an important reason for preventing development.

Although absolute devolution of ownership and management of the New Forest has never been granted, the Court of Verderers represents a form of devolution as a statutorily configured but partly locally elected decision-making body with an effective power of veto over other statutory bodies. Since 1877, the New Forest Acts (1877, 1949, 1964, and 1970) have not only defined the powers and duties of the Court of Verderers, but also circumscribed the activities of all the other collective bodies with management interests in the Crown lands, including the Forestry Commission, the commoners, and the general public. The Forestry Commission must seek the agreement of the Verderers in order to "provide, manage, maintain and improve tourist, recreational or sporting facilities and any ancillary equipment, facilities or works" and consult with them before carrying out "all such silvicultural maintenance works as may from time to time be necessary" (New Forest Act 1949, section 18; New Forest Act 1970, section 1; Forestry Act 1967; Countryside Act 1968). For many decades, commentators and politicians have referred to this relationship of shared power between the Forestry Commission and Court of Verderers as a system of "checks and balances".

The Verderers' Court still meets monthly in the Verderers' Hall (circa 1297). A variety of people attend the Court, including representatives of the New Forest-specific interest groups, retired commoners, and other interested individuals. The timing of the court, however, prevents most of the commoners in full-time employment from attending. Typically, the court attracts around 40 to 50 members of the public, depending on the time of the year. Far from being an inclusive, modern, and democratic debating arena, the ancient court follows a very traditional, formal procedure, being opened by the *Senior Agister* with a pledge of allegiance to the Monarch. "Presentments" are made from the dock and comprise requests for assistance from the Verderers by commoners, other governing bodies, or any individual with an interest in the New Forest. After the public session is completed, the Court continues its discussions in private and decisions of the Verderers are not announced until the next month's session.

Although the Verderers' Court was reconstituted as a forum for resolving problems affecting common forest rights, it is regularly employed by residents who appeal to the Verderers to use their power of veto over the Forestry Commission to ensure that no unwanted recreational development takes place in the New Forest, and to use their influence to protect the New Forest from other such change. As such, local residents are vociferous defendants of the Verderers' existence and strongly supported the call for a special Act of Parliament to constitute the New Forest National Park at its inception in 2004. Instead, the government created a New Forest National Park using the national generic legislation contained in the Environment Act 1995, but promised to protect the institutional arrangements that had evolved over 1000 years. Ironically, if the government had been prepared to make parliamentary time to introduce the tailor-made legislation called for, it is quite likely that it would have used the opportunity to review, reform, and even abolish the ancient Court of Verderers: the court's exclusive voting system has been at odds with England's principles of democracy for over a century.

It has become accepted that residents with no formal rights to the Open Forest will use the New Forest commoners' forum, the Court of Verderers, to defend their own interests. However, the events of 2008 marked a sinister turn in such rent-seeking behavior, as third-party activists appeared to hijack the commoners' very persona.

The Recent "Revolt" and the Rights Impostors

The so-called revolt of September 2008 was the result of a challenge to dog walkers and horse riders in the New Forest by the New Forest National Park Authority (NFNPA). Both types of users were criticized in the plan: the horse-keepers for dividing up fields into smaller paddocks, building shelters, and stables, and generally converting a rural, pastoral landscape into a "horsiculture" landscape; the dog walkers for disturbing ground-nesting birds during the breeding season and so threatening an important part of the New Forest's ecology. The NFNPA hinted that it would be looking for tighter restrictions on subdivision of fields into smaller paddocks and exploring the merits of closing some of the 150 New Forest car parks in order to redirect dog walkers away from vulnerable ecological sites.

Several large meetings were convened and the support of the two local members of parliament called upon to prevent any such restrictions appearing in the management plan. In its publicity, the Forest Uprising group declared that "the peasants are revolting", implying their members comprised the original New Forest commoners and aligning themselves with the 1000 years of oppression from the Crown and the state. Newspaper correspondence reflected the discourse of an oppressed peasant stock: "Those who live in the New Forest have been effectively disenfranchised. The creators of the Magna Carta will be spinning in their graves to see how weak and soft modern English people are to have just handed away their right to self determination without a fight" (Bell, 2009). An eagerness to align themselves to commoners' interests caused a rush of amenity users to apply to join the Commoners Defence Association.

Nowadays, the term "commoner" is usually reserved for those not only possessing rights, but also practicing them. Around 500 commoners exercise grazing rights, turning out around 7000 animals each year (NFCDA, 2010). Commoning is now almost exclusively a part-time occupation: some 90 percent of commoners farm no more than 20 acres (80ha) and over 80 percent turn out 20 animals or less. Commoners cooperate, provide mutual support for one another, and share knowledge and experience through pony drifts, auctions, markets, and other meetings, sharing the New Forest pony as an essential link: "the Forest pony in a humbler more democratic way is a tribal god among the commoners" (Kenchington, 1944, p173). In truth, most horse keepers and dog walkers comprise stakeholders who have relatively recent presence in the New Forest, and few have claims in terms of defined rights to the Crown land. Indeed, the amenity interest in the New Forest, including nature conservation and recreation, is often at odds with the commoners' interests. Both dog walking and picnicking are seen by the commoners as incompatible with grazing livestock when enjoyed by large numbers of visitors, because of the disturbance to stock by people who invariably attempt to feed the ponies, and their dogs, which sometimes chase them. As for the horse riders, the majority are just that, *horse* riders: only a few ride New Forest ponies and so contribute to the local economy of pony breeding; the vast majority choose larger thoroughbred crosses on which to enjoy the forest.

The strongly defended interests of the local stakeholders proved a formidable obstacle to change. Following the resignation of the New Forest National Park Authority's chief executive, a revised plan was issued, with notably watered down proposals relating to restrictions on dog walking and horse keeping. For the commoners, these "rights impostors" marked yet another example in history of other stakeholders upholding common forest rights as a means to protect their *own* interests in the New Forest. It is interesting to note the parallels with traditional forest communities in the Amazon, where "the new allies pursued their own interests and dominate the partnership" (Medina et al, 2009), and with communities in transition in California (Walker and Fortmann, 2003). Wise to the self-interested motives of new applicants, the Commoners Defence Association immediately closed its doors to new members during the "revolt".

Conclusions: Lessons Learned for the Forest Rights Agenda

This book is about the rights agenda in international forestry. Chapter 1 explained how that agenda demonstrates a strong orientation towards social justice, with activists calling for equitable distribution of forest benefits, advocating the recognition of forest peoples' culture and promoting the participation of forest people in political decision-making. In the New Forest's history, there is little documented evidence of the part that the commoners as indigenous rights-holders played in early contests against the monarchy and nobility; but there is general agreement that they were receivers rather than participants in institutional change. Throughout most of the New Forest's history, the commoners were poorly educated, geographically dispersed, and fiercely individualistic in nature, and lacked the social capacity and experience to engage meaningfully in institutional debates. Indeed, the history of New Forest activism has been one of sovereign rule, checked and contested by the self-interested "rent-seeking" behavior of the more powerful local inhabitants. Despite a growing rhetoric of "participatory" governance, research suggests that certain groups continue to dominate the decision-making practices of community forestry around the world (see, for example, Ojha et al, 2009, and Oyono, 2005). The arrival during the 19th century of a middle class in the New Forest broke a social dichotomy between the wealthy land barons and the commoners. However, despite claiming to defend the commoners' rights, the new middle-class protagonists have been as self-seeking as their aristocratic predecessors, often employing the commoners' rights to protect the integrity of the New Forest for their own interests.

As different kinds of social actors assert competing claims, policy-makers are forced to make difficult choices about whose claims to support. In doing so, they need to be aware of how the existing institutional arrangements have been constructed and during which political pressures. Historical institutions can carry with them strong biases that reflect historical favoritism towards specific users, and social or ethnic groups. A fundamental problem in the New Forest is that its system of "checks and balances" is the result of a long history of contests of self-interest and self-preservation. It is far from proven that the requirements of different stakeholders, when demanded in aggregate, can be met completely within the sustainable capacity of the New Forest. If such needs cannot be met completely, then a debate about distribution and fairness takes over. The decision-making arenas in the New Forest remain crucial in terms of their ability to influence allocation decisions. However, equal access to such arenas continues to remain elusive. The so-called "revolt" of 2008 showed that even after 1000 years, the local amenity stakeholders were willing to claim *de facto* rights to the Open Forest and imitate the commoners' cultural identity in order to further their own interests. Eventually, the commoners of the New Forest began to assert themselves in defending their ancient common rights by closing entry to the Commoners Defence Association. However, they still have a long way to go in claiming their proper role in decision-making arenas, particularly the Verderers' Court.

References

Bell, R. (2009) "National park: Letter to the editor", *Lymington Times*, 25 April 2009, p23

Cobett, W. (1830) *Rural Rides*, London

Hutchinson, H. G. (1907) *The New Forest*, third edition, Methuen & Co, London

Kenchington, F. E. (1944) *The Commoners' New Forest*, Hutchinson, London

Medina, G., B. Pokorny, and J. Weigelt (2009) "The power of discourse: Hard lessons for traditional forest communities in the Amazon", *Forest Policy and Economics*, doi:10.1016/j.forpol.2008.11.004

Mudie, R. (1838) *Hampshire*, Gilmour, Winchester, UK

NFCDA (New Forest Commoners Defence Association) (2010) *Annual Report and Statement of Accounts for 2009*, NFCDA, New Forest

NFNPA (New Forest National Park Authority) (2008) *New Forest National Park Plan Consultation Draft: National Park Management Plan and Local Development Framework Core Strategy and Development Policies*, NFNPA, Everton, Lymington, UK

Ojha, H. R., J. Cameron, and C. Kumar (2009) "Deliberation or symbolic violence? The governance of community forestry in Nepal", *Forest Policy and Economics*, doi:10.1016/j.forpol.2008.11.003

Oyono, P. R. (2005) "Profiling local-level outcomes of environmental decentralizations: The case of Cameroon's forests in the Congo Basin", *The Journal of Environment and Development*, vol 14, no 3, pp317–337

Pasmore, A. (1977) *Verderers of the New Forest: A History of the New Forest 1877–1977*, Pioneer Publications, Beaulieu, UK

Stagg, D. (1984) *Verderers of the New Forest*, New Forest Association, UK

Steins, N. A. and V. M. Edwards (1999a) "Collective action in common-pool management: The contribution of a social constructivist perspective to existing theory", *Society and Natural Resources*, vol 12, no 539–557

Steins, N. A. and V. M. Edwards (1999b) "Platforms for collective action in multiple-use common pool resource management – Special Issue", *Agriculture and Human Values*, vol 16, pp241–255

Tubbs, C. R. (1986) *The New Forest*, Collins, London

Vesey-FitzGerald, B. (1977) *Portrait of the New Forest*, second edition, Robert Hale, London

Walker, P. and L. Fortmann (2003) "Whose landscape? A political ecology of the 'exurban' Sierra", *Cultural Geographies*, vol 10, pp469–491

Wise, J. R. (1863) *The New Forest, its History and Scenery*, second edition, Smith, Elder, and Co, London

10

Advocating for Traditional Native American Gathering Rights on US Forest Service Lands

Beth Rose Middleton

We don't want to be made into criminals again, but we will not stop gathering just because someone in Washington, DC, thinks they know better than we do how to sustain traditional plants.[1]

The spot where the tea grows is just off the road, up a small hill, in a seep below where a village once perched over the valley. I accompany a friend, a Maidu traditionalist, to the site, and at other times to gather willows for basketry on private land. He explains to me that this particular tea-gathering place (on Forest Service land), and the willow site (on private land), and another place where we will go later for maple – again on Forest Service land – and yet another location where we will harvest stick tea (on Bureau of Land Management Land) all form a series of places where his relatives have been gathering for centuries. The jurisdiction of the land has changed over time; yet, the identity of the gathering places, the need to return to them to maintain the plants; and the knowledge among other families that this is his family's place to gather all remain. As we fill bags of elderberries and toss coins as offerings into the bushes, I reflect on numerous other contributions that may lie there beneath the duff: a history of acknowledgement for the harvest. As we sing to the plants and carefully gather,

I think about how the plants themselves are as much relatives as the ancestors who were here tending them just decades before.

Fluctuating Rules of Law

Within Sikor and Stahl's framework of activism to recognize forest people's rights (see Chapter 1), such rights are often overlooked by agency decision-makers who do not recognize the needs of culturally distinct, indigenous forest-dependent populations. For example, even as the traditional stewardship practices noted above continue, the agencies overseeing public lands are continually altering their management approaches, sometimes with disastrous consequences for traditionalists.

One such fluctuating area of Forest Service management regards revenue generation from forest products. While the agency's primary revenue has historically come from timber, it has long recognized the potential for marketing other forest products. As early as 1928, a *National Forest Manual* instructed regional foresters to develop guidelines for the sale and valuation of non-timber forest products (McLain and Jones, 2005, pp5–6). By 1949, the Forest Service was collecting fees for the harvest of Christmas trees, ferns, and other species. Following the 1980s/1990s decline in the timber economy, the Forest Service released *Income Opportunities for Special Forest Products: Self-Help Suggestions for Rural Entrepreneurs* (Thomas and Schumann, 1993). Special forest products harvesting is decentralized, making it a feasible industry for rural areas; but the lion's share of the profits do not go to the local producer. Rather, value increases as the product changes hands, tending to accrue with the end processor, depending on the specific product. As more rural people and contracted laborers began mushroom harvesting, *Echinacea* digging, and salal picking, reports of increased revenue began coming from the non-timber forest products industry.[2] With multi-million dollar economies emerging for matsutake and other mushroom species, floral greens such as salal, herbal medicines such as goldenseal, and landscaping materials such as pine straw, the Forest Service began selling permits for special forest products harvesting.

In 2000, the Department of Interior and Related Agencies Appropriation Act included a Pilot Program of Charges and Fees for Harvest of Forest Botanical Products.[3] The program, set to last through 2004, scheduled the US Forest Service to charge gatherers for non-timber forest products, and then to apply the revenue to operate the oversight program, including ecological monitoring. In 2004, the deputy chief of the National Forest System directed all forest supervisors to consult with federally recognized tribes regarding future special forest products legislation, following the recommendation of an assessment team that such legislation would have implications for tribes. In 2007, during the second George W. Bush administration, a draft system-wide gathering policy was published in the federal register. Some tribes, particularly unrecognized tribes, were not aware of the proposed rule until this publication. Tribes were largely opposed to the policy, which lumps them in with commercial gatherers; requires disclosure of detailed information, including the location, frequency, and quantity of materials gathered; and, in some cases, charges for permits.

Following the 2007 publication of the draft rule, California Indian Basketweavers' Association (CIBA) policy analyst Jennifer Kalt circulated information about the rule and what it would mean for tribes.[4] Many tribal members found the policy insulting and in violation of their rights under the American Indian Religious Freedom Act, the National Historic Preservation Act, the Executive Order on Environmental Justice, and other statutes. Like many US government policies towards Indians, the policy also sought to divide people within Indian country: treaty tribes' rights were affirmed (but limited), while tribes who never negotiated treaties, or whose treaties were never ratified, were to be penalized by their lack of a treaty – they would have no more rights to gather on Forest Service land than a commercial harvester.

Despite organized opposition led by CIBA, on 29 December 2008, in the last hours of the Bush administration, the final rule on the Sale and Disposal of Special Forest Products and Forest Botanical Products was published in the federal register, and set to take effect on 28 January 2009.[5] The rule addressed very different populations with different interests in the forest: large companies that subcontract to forest laborers; self-employed forest products harvesters; and traditional gatherers maintaining resources for continued, long-term cultural use. On 20 January 2009, a memo from former White House Chief of Staff Rahm Emanuel delayed all regulations that had not yet been reviewed by the Obama administration, extending the public comment period until 2 March 2009, and prompting a flurry of additional comments from tribes and their allies. This chapter explores some of the ways in which tribal members, Native advocacy groups, and allies responded to this policy by articulating indigenous stewardship and gathering practices as rights on US Forest Service lands, particularly in California.

California: How tribal lands became US Forest Service lands

While tribes throughout the US were negotiating treaties with the federal government to cede certain lands, retain others, and underscore their usufructory rights to traditional territories, the treaties made with California tribes in 1851 and 1852 were buried in the Senate archives for 50 years and never ratified (see Heizer, 1974; Hoopes, 1975; and Prucha, 1994). As such, California tribal leaders negotiated away their homelands in exchange for protection and reserved rights that never materialized. Today, California has the largest population of Native Americans in the US; but there are no tribes with treaties with the federal government. In addition, many federally recognized California tribes have very small land bases, some established through the appropriation of funds between 1914 and 1924 to purchase lands for "homeless California Indians".[6]

While tribes were being dispossessed of their lands following the treaty-making period, individual settlers and large companies were buying these same lands, and cutting great stands of timber, damming rivers, re-routing creeks, and draining and tilling marshes and meadows. Realizing that forest reserves throughout the US might soon be exhausted, the federal government introduced the Forest Reserve Act in 1891, setting aside certain forestlands from purchase and development. The lands that became the Forest Reserve (and, in 1905, US Forest Service lands) could legally be incorporated within the public domain only after Indian title had been extinguished. Indian title was never extinguished for much of these lands in California; but effectively ignoring it and committing genocide against Indian people served to remove many Native people from the Forest Reserve.[7]

Because of their lack of reserved tribal homelands, many tribes kept their cultures alive by continuing to steward the natural and cultural resources on lands removed from their jurisdiction. Tribal members often had to maintain cultural practices in secret because of assimilationist laws outlawing Indian religious practice, and forced enrollment of Indian children in Indian boarding schools, where traditional cultural knowledge and the speaking of native languages were prohibited. Not only did it take courage to continue traditions in this environment, it was also very difficult to pass them on, as families were separated, and children were orphaned by disease and violence.

Those maintaining patches of basketry, medicinal, and other resources also risked prosecution by government landowners for using agency resources, although those resources had been under tribal control until just 50 years prior to the establishment of the Forest Service, and were, in effect, still tribal lands. Despite this logic, legally the lands had become Forest Service lands, subject to agency restrictions and oversight. The Forest Service operates on principles of multiple use and conservation, with varying emphasis on different activities depending upon each successive administration. What cannot change between administrations, however, is the related trust responsibility that the US government has to American Indian people. In exchange for some relinquishment of their lands and freedoms, Native Americans are guaranteed that lands and resources important for their survival are held in trust for them by the agencies of the US government. In addition, laws are to be interpreted in a way that is favorable to Native interests. Tribes' reaction to the proposed final rule is one example of pushing the US Forest Service to uphold its trust responsibility by ensuring that resources important to the continuation of Native culture are protected and accessible to Native Americans.

Of course, this trust responsibility only officially applies to federally recognized tribes, which have had their status as aboriginal governments affirmed by the US government. There are many Native people who, due to lack of ratification of treaties, assimilationist policies, and violent warfare against tribes, were never recognized as distinct governments, or had their governmental status terminated. According to research by the American Indian Policy Review Commission, California has the largest group of federally unrecognized tribes in the US.[8] Over 55 tribes in California are unrecognized out of approximately 133 unrecognized tribes throughout the nation. The number of unrecognized tribes and individuals in California dates directly to the non-ratification of California Indian treaties.

This disparity for California tribes was underscored by the proposed rule, which exempted treaty tribes from permitting requirements to gather non-timber forest products, creating a system of preferential treatment for tribes whose treaties had been ratified. As the California Indian Basketweavers' Association responded: "California Indians should not be treated differently from other tribes just because the federal government failed to live up to its promises in the past" (Parker and Kalt, 2008). The proposed final rule also noted that even treaty-protected gathering rights would be subject to Forest Service regulation, releasing a firestorm of comments from tribes protecting their treaty rights, and weakening the rule. The following sections will discuss tribal members' observations of non-timber forest products harvesting, and then trace some principles of the rule and indigenous rights-based objections. The chapter will conclude with a summary of where the rule is today, and with reflections on the power of a Native American rights-based forestry agenda.

"The Forest Was Our Backyard"

As the special forest products industry grew, tribal members who had long been stewarding and harvesting particular resources on national forest system lands within their traditional territories began to notice a decline. As Renee Stauffer (Karuk) recalled, when she returned to her hometown of Orleans during the late 1980s, a boom in forest products harvesting was occurring, and she encountered commercial gatherers on the land. "The forest was our backyard", she explained. "We considered it our land."[9] Indeed, the Karuk Tribe's government-to-government status was restored in 1979, but ancestral lands were not returned.[10] As such, the Karuk Tribe's agreements with the surrounding Six Rivers National Forest enable their stewardship of sites that family members have been tending for centuries.

While some species that are important to the tribes may not be important to commercial harvesters, others, such as beargrass, tanoak mushrooms, and huckleberries, are sought after by both, creating conflict. For many traditionalists, the Forest Service added insult to injury by promulgating a rule regulating both tribal and commercial gathering, when tribal gathering has not depleted the resource, and tribal traditions are stressed by commercial gathering. Some Native commenters expressed their outrage at this problematic aspect of the rule:

> *Our tribal sovereignty and the government to government with the USFS [US Forest Service] should be left alone. Indians have been taking care of the forest way before that dumb ass Columbus got lost and ended up on our doorstep. You need to worry about the non-Indians who take to their hearts content for a profit and*

no concern about over-harvesting areas. Maybe you should get a permit to go grocery shopping and get your medicine. Leave our traditions alone.[11]

Proposed Special Forest Products Rule

The proposed rule sought to establish sustainable harvest levels for non-timber forest products, and to generate revenue from the regulation of these products to pay for their oversight and restoration. The rule attempts to establish monitoring and enforcement protocols for both special forest products (including firewood, posts and poles, wildflowers, fungi, moss, nuts, seeds, and Christmas trees), and the sub-category of forest botanical products which excludes any timber products, such as Christmas trees and firewood.[12] The agency proposes to establish personal-use harvest levels for each forest botanical product. Any use below that level will be free, and any products obtained for free cannot be resold. In order to gather products above one's personal harvest level, the gatherer must purchase a permit, and note where they will be gathering, how much, how often, and other information. The agency recognizes "treaty and other reserved rights", but reserves authority to set harvesting levels and prohibit commercial sale, and does not recognize the rights of non-treaty tribes or federally unrecognized tribes.

The need for regulation

By and large, Native people commenting on the rule did not oppose some regulation of the commercial forest products industry; but they wanted recognition that their rights to the land are different – steeped in tradition, survival, and distinct political–legal precedent. As one US Forest Service employee summarized:

> *Most tribal people do want some type of regulation on the non-native gatherer that is out there impacting these resources ... If the proposed rule removed the personal use, non-commercial component it was ... a decent policy to deal with the non-Native commercial gatherer, but [it made the] mistake of lumping the commercial gatherer with the Native together ... that is my own opinion. These two need to be separated.*[13]

In attempting to document Native concerns with the proposed final rule, I coded a sample of 123 comments submitted between 2007 and 2009 by tribes, tribal members, tribal lawyers, and allies regarding the proposed and final versions of the special forest products rule. Of this sample, 122 comments specifically indicated a need for regulation of special forest products harvesting, and 120 noted that traditional harvesting was categorically different from commercial harvesting. As Alicia Cordero (Chumash) wrote in her comment:

> *For historical reasons of which I'm sure we are all aware, the native people of this land are few in number. Within this small number of people, only a fraction of us utilize the natural materials within the national forests. The impact of our gathering is small. Unfortunately, there is a non-native population that harvests unsustainable amounts of materials, such as white sage, for profit, viewing hillsides of sacred plants as mountains of free cash. These people in no way represent the collecting practices of native people. Clearly, some sort of regulation and enforcement is necessary to prevent this kind of abuse.* (Cordero, 2009)

Articulating opposition to the rule

Not only was it an affront to be lumped in with commercial gatherers, but the rule also did not recognize traditional gatherers as forest managers contributing to overall forest health (see, for example, Kimmerer and Lake, 2001; Anderson, 2005; Cunningham, 2005; Hankins, 2005; Lightfoot and Parrish, 2009). As Tuolumne Me-Wuk Chairman Kevin A. Day wrote in his 2007 comments:

> *... gathering has been part of the ecosystem – removing traditional gathering could have adverse effect on the environment. Have you evaluated the potential of this? What will be the effect of this rule on the health and monitoring of the forest?* (Day, 2007)

Non-treaty traditional gatherers would also have to comply with personal-use harvest levels that differ from traditional gathering quantities. As Vicki Stone, also of Tuolumne Mi-Wuk, asked in her 2009 comments:

> *How much is considered personal use? Who decides how much is personal use? Can I only gather enough acorns that I can eat instead of gathering for the elders of our tribe? Acorns are not gathered on a personal use harvest level.* (Stone, 2009)

Non-treaty gatherers would also have to identify locations of use, which is highly problematic for reasons of both cultural privacy and the safety of the resources. As traditional basket weaver LaVerne Glaze (Karuk) remembered, she and her daughters initially tried to protect their mushroom-harvesting areas from commercial harvesters by posting Forest Service signs prohibiting commercial harvest and denoting traditional Karuk mushroom-gathering areas. "That backfired", she explained in 2010; "people sought out those spots to gather".[14] Reporting gathering locations to the Forest Service makes them highly vulnerable to pillage by other users. As explained in comments from California Native sacred sites advocacy group, the United Coalition to Protect Panhe:

> *Permits should not be required for traditional gathering for basketweaving, medicinal, ceremonial, or subsistence food uses. Requiring permits invades traditional cultural practitioners' privacy, since information submitted on permit applications can be made public. Making public which plants are used or gathering area locations can threaten plant populations and traditional practices.* (Robles and D'Arcy, 2009)

Culturally, gathering locations are often kept private. Basket weavers do not want to be watched or otherwise disturbed while they are saying prayers and singing gathering songs, and even disclosing a gathering spot can adversely affect the efficacy of the plants.[15] Tribal members argue that site disclosure violates their rights under American Indian Religious Freedom Act, the Religious Freedom Restoration Act, and the First Amendment of the Constitution (Nuvamsa, 2008).

National policy-makers can learn from regional successes

The proposed rule also flies in the face of successful regional policies and memoranda of understanding (MOUs). In comments regarding the rule, the Bureau of Indian Affairs underscored the importance of actualizing the federal government's trust responsibility through effective policy-making:

> *Many of these lands were included in Indian Treaties that, due to circumstances associated with the California gold rush, were never given to the tribes in California. Subsequently, some tribes in California were placed on Reservations and Rancherias that are less than one acre in size. This makes it very difficult for Indian people who traditionally utilized plant materials collected from a large land base for cultural, spiritual, and medicinal purposes to find suitable plant material on Indian trust lands. The existing Regional Policy combined with PRO [Pacific Regional Office] gathering policies assists our agencies in ensuring sustainability of the resources and [ensures] that our Federal obligation to Indian Beneficiaries is met.*[16]

The regional policy referred to is the US Forest Service Pacific Southwestern Region (Region 5) Traditional Gathering Policy, developed by representatives from CIBA, the California Indian Forest and Fire Management Council (CIFFMC), the Bureau of Land Management (Department of the Interior), and the Forest Service (Department of Agriculture). Regional Tribal Relations Program Manager Merv George, Jr., of the US Forest Service, was active in the creation of this policy in his previous role as executive director of CIFFMC. He described participating in years of meetings, representing tribal natural resource issues along with other Native entities, and working with agency representatives to develop guidelines that respect traditional gathering rights. He calls the resulting policy "a model for collaboration between our tribal communities and the federal land management agencies".[17]

Signed in 2007, the policy specifically addresses both traditional access to plants and the management of these plants to ensure their continued abundance. The policy acknowledges that recognized and unrecognized tribes rely on the resources on these lands, notes that traditional gathering and stewardship benefit the ecosystem, and affirms that "personal use should have preference over commercial use". The US Forest Service and Bureau of Land Management regions party to this policy clearly offer their "support" to "traditional native cultural practitioners [federally recognized or unrecognized] in gathering culturally utilized plants".[18] The policy directs agency field offices to work with local Native stakeholders to create specific gathering policies, and to collaborate with tribes on restoration. In sum, the policy clearly recognizes that local indigenous peoples have rights to the forest, and that these rights should be respected, supported, and enhanced.

However, since the Traditional Gathering Policy does not impose fees or permits, or call for disclosure of gathering sites, it is inconsistent with the proposed rule and would be declared invalid if the rule were affirmed. According to Kalt, this would do severe damage to US Forest Service tribal relations: "essentially it would undercut the word of local USFS staff, making future negotiations difficult since they might be seen as untrustworthy".[19]

Current Status

In response to a preponderance of comments on the rule's non-compliance with treaty rights, confusion of traditional gathering with commercial gathering, violation of the federal trust responsibility, violation of the American Indian Religious Freedom Act and other statutes, and contradiction with the overriding "Farm Bill" (HR 6124), the proposed final rule was officially delayed indefinitely on 1 June 2009.[20] Traditional gatherers in California breathed a collective sigh of relief that their existing regional Traditional Gathering Policy would be upheld. CIBA and other groups continue to watch the register for any resurgence of the rule. In addition, many gatherers, tribes, Native non-profit organizations, and tribal allies made proactive suggestions for an improved rule that placed traditional use over personal and commercial use; recommended that fee and permit waivers be granted to all tribes and to tribal governments; and offered their expertise to develop sustainable harvesting guidelines for traditionally important species.[21]

Rights-Based Agenda

These calls for participation in rule-making, and for modification of the rule to meet tribal priorities, were consistently articulated in a language of rights.

Tribes argued for recognition of their traditional rights held prior to European contact:

> *As a federally recognized tribe, the Pit River Tribe values land in its ancestral territory that hold cultural values and meaning, and are still actively used today to assure the cultural and spiritual continuity of the Tribe. Therefore, this system of permits and/or fees for traditional gathering ("personal use") would subordinate the natural and spiritual characteristics of the Tribe ... The Pit River Tribe* strongly *supports and recommends adding language to this policy that would include permit waivers for all traditional use by American Indians.* (Villarruel, 2009)

In further comments, the Pit River Tribe also asks the Forest Service to develop a different standard for traditional tribal commercial use. Tribes have long depended on these same forestlands for subsistence, and on revenues derived from the creation and sale of baskets, jewelry, and other forest-based products. This traditional economy has not depleted forest resources. Tribes argue that not only is it their right to continue to access their ancestral lands for subsistence and survival, but that their resources need to be safeguarded and stewarded.

Indigenous lands were taken from Native Americans during a period of time when discussion of indigenous rights was out of the question. In California, in particular, treaties negotiated with indigenous peoples were never ratified, leaving California tribes without treaty-based rights. However, contemporary Native opposition to the proposed forest products rule is empowered by rights-based frameworks as the 2007 United Nations Declaration on the Rights of Indigenous Peoples. The latter affirms the rights of indigenous peoples to traditional homelands, cultural practices, and distinct cultural-political communities.[22] Actions that would attempt to assimilate indigenous peoples by forcing people to stop specific cultural or spiritual practices, or denying access to traditional homelands, are prohibited. Drawing on this language, as well as the comments from tribes, traditional gatherers, and Native non-profits, the next version of the special forest products gathering rule should recognize the unique roles, needs, and expertise of indigenous gatherers (recognized and unrecognized; treaty and non-treaty) and explicitly acknowledge their *rights* to American forests.

Notes

1 Anonymous basket weaver (cited in Kalt, pers comm, 28 August 2009). A similar sentiment was expressed by Karuk basket weaver Renee Stauffer: "[The rule will] just turn a bunch of innocent basket weavers into criminals. It's not going to stop us from gathering. Baskets are just a part of our lives" (cited in Walters, 2009).

2 See, for example, Schlosser et al (1991); and McLain et al (2008). For information on the broad demographic of harvesters (including Anglo back-to-the-landers and/

or loggers out of work, Latino or Southeast Asian immigrants, and rural Native Americans), see Solberg (1999); Brown and Marin-Hernandez (2000); and Emery et al (2003). For examples of studies on non-timber forest products, see Bagby (2004) and McLain et al (2005) on mushroom harvesting; Solberg (1999) on *Echinacea*; and Lynch and McLain (2003) and Ballard and Huntsinger (2006) regarding salal.

3 See *News & Views* (2000, p5).

4 CIBA is a traditional basket weavers' advocacy group formed in 1989. According to a history of CIBA provided by the organization, research on the challenges facing California Indian basket weavers began in 1986, a group of basket weavers first gathered in 1989, and CIBA was officially incorporated in 1992 (CIBA, undated).

5 Federal Register 73(249): Monday, 29 December 2008. 36 CFR Parts 223 and 261, "Sale and Disposal of National Forest System Timber; Special Forest Products and Forest Botanical Products".

6 This series of appropriations began in 1914 with 38 Stat. 582–589 and culminated in the 1922 "Purchase of Land for Homeless Indians of California", 42 Stat. 559–567. In 1923, the California State Assembly also passed A. B. 1333 to set aside 1488 lands for the tribes in Plumas, Lassen, and Modoc Counties, northeast California.

7 Individual Indian people still advocated for their lands within the boundaries of the US Forest Service. An Act of 25 June 1910 (36 Stat. L. 855) authorized Indians living on land within the Forest Reserve to apply for allotments.

8 The American Indian Policy Review Commission (AIPRC) was created in 1975 (Public Law 93–580) and charged with examining the administrative and legal relations between American Indians and the federal government. Data on California is from a 1977 AIPRC report, analyzed in Goldberg and Champagne (1996).

9 Pers comm, Renee Stauffer, Eureka, CA, 15 February 2010.

10 A 15 January 1979 memorandum from the Assistant Secretary of Indian Affairs, entitled "Revitalization of the Government-to-Government Relationship between the Karok [sic] Tribe of California and the Federal Government" confirmed that the continued existence of the Karuk Tribe had been "substantiated" and recommended re-establishing the government-to-government relationship.

11 Comment on 36 CFR Part 223, sent to the US Forest Service via electronic communication, 24 February 2009.

12 See US Forest Service (2009) *Special Forest Products Focus of Final Rule*, 1 September 2009, www.fs.fed.us/fstoday/090109/02National%20News/special.html; and Federal Register 72(203), 22 October 2007, Proposed Rules, Subpart H – Forest Botanical Products, § 223.277 Definitions.

13 Pers comm, US Forest Service employee, 2 February 2010.

14 Pers comm, LaVerne Glaze, Eureka, California, 15 February 2010

15 CIBA Policy Analyst Jennifer Kalt shared that several basket weavers had expressed these sentiments to her; electronic communication, 22 March 2010.

16 Letter from the acting regional director of the US Bureau of Indian Affairs to the forest management director, USDA Forest Service, entitled "Comments on proposed rule, sale, and disposal of national forest system timber; special forest products and forest botanical products", 20 December 2007.

17 Pers comm, Merv George, 22 March 2010.

18 *Forest Service Manual Supplement*, no 1560, Region 5, approved July 2007; Bureau of Land Management Instruction Memo no CA-2007-017, approved April 2007.

For a shortened version of the Traditional Gathering Policy, see the website of the California Indian Basketweavers' Association at www.ciba.org/TraditionalBrochure.pdf; for some background on the policy and the full text, see Bernie Weingardt (regional forester, Pacific Southwest Region, Forest Service) and Mike Pool (California State director, Bureau of Land Management), Memo to area managers and forest supervisors, 29 November 2006.

19 Electronic communication, Jennifer Kalt, 22 March 2010.

20 The 2008 Farm Bill, Title VIII, Cultural and Heritage Cooperation Authority, Subtitle B, allows for confidentiality of cultural sites and practices (Section 8106), and provides for free use of forest products for traditional and cultural uses (Section 8105).

21 See, for example, comments from Ortiz (2008) recommending that the proposed rule be modified to "consider American Indian traditional gathering separately from other types of personal use, and give primacy to personal use over commercial use"; and the Intertribal Timber Council (2008), supporting a "separate tribal standard in recognition of the inherent cultural, ceremonial, and/or traditional interests tribes maintain on forest lands now part of the National Forest system". The California Native American Heritage Commission (NAHC) "strongly recommend[s] that the new rule ensure access without fees or permits to all tribes and tribal members for gathering plants of importance for traditional, ceremonial, and/or cultural uses as protected under the American Indian Religious Freedom Act", and San Manuel Band of Serrano Mission Indians Chairman James Ramos (2009) "request[s] that the Department of Agriculture, Forest Service (USDA/USFS) exempt all Native Americans regardless of whether they are federally-recognized and/or treaty tribes".

22 *United Nations Declaration on the Rights of Indigenous Peoples* (2007), particularly Articles 5, 8, 10, 11, 24–26, and 29.

References

Acting Regional Director, US Bureau of Indian Affairs (2007) "Comments on proposed rule, sale, and disposal of national forest system timber; special forest products and forest botanical products", 20 December 2007

Anderson, M. K. (2005) *Tending the Wild: Native American Knowledge and the Management of California's Natural Resources*, University of California Press, Berkeley, CA

Assistant Secretary of Indian Affairs (1979) "Revitalization of the government-to-government relationship between the Karok [sic] Tribe of California and the federal government", 15 January 1979

Bagby, K. (2004) *Sharing Stewardship of the Harvest: Report on the 2003 Crescent Lake Mushroom Monitoring Project*, June, Pacific West Community Forestry Center, Taylorsville, CA

Ballard, H. L. and L. Huntsinger (2006) "Salal harvester local ecological knowledge, harvest practices and understory management on the Olympic Peninsula, Washington", *Human Ecology*, vol 34, no 4, pp529–547

Brown, B. and A. Marin-Hernandez (eds) (2000) *Voices from the Woods: Lives and Experiences of Non-Timber Forest Workers*, Jefferson Center for Education and Research, Wolf Creek, OR

Bureau of Land Management (2007) *Bureau of Land Management Instruction Memo*, no CA-2007-017, approved April 2007

CIBA (California Indian Basketweavers' Association) (undated) *A Brief History*, obtained from Jennifer Kalt, former CIBA policy analyst; on file with author

Cordero, A. (2009) Comment letter sent to the US Forest Service entitled "Re: New United States Forest Service Final Rule (36 CFR Parts 223 and 261): Sale and disposal of national forest system timber: Special forest products and forest botanical products", 1 March

Cunningham, F. (2005) "Take care of the land and the land will take care of you: Traditional ecology in Native California", *News from Native California*, vol 18, no 4, summer, pp24–34

Day, K. A (2007) Letter on behalf of Tuolumne Mi-wuk Tribal Council to the director of Forest Management, USDA Forest Service, regarding "Forest rule on the sale and disposal of national forest system timber; special forest products; and forest botanical products", 20 December

Emery, M. et al (2003) "Special forest products in context: Gatherers and gathering in the Eastern United States", USDA Forest Service Northeastern Research Station, GTR-NE 306

Federal Register (2008) Federal Register 73(249): Monday, 29 December 2008, 36 CFR Parts 223 and 261, "Sale and disposal of national forest system timber; special forest products and forest botanical products"

Forest Service (2007) *Forest Service Manual Supplement*, no 1560, Region 5, approved July 2007

Goldberg, C. and D. Champagne (1996) *A Second Century of Dishonor: Federal Inequities and California Tribes, Chapter I*, Report prepared for the Advisory Council on California Indian Policy, www.aisc.ucla.edu/ca/Tribes1.htm

Hankins, D. (2005) *Pyrogeography in Riparian Ecosystems: An Indigenous Tool for Resource Management and Conservation*, PhD thesis, University of California, Davis, CA

Heizer, R. F. (1974) *The Destruction of California Indians*, University of Nebraska Press, Lincoln, NB, republished 1993

Hoopes, C. (1975) *Domesticate or Exterminate: California Indian Treaties Unratified and Made Secret in 1852*, Redwood Coast Publications, Loleta, CA

Intertribal Timber Council (2008) "Re: Proposed rule, request for comment on sale and disposal of national forest system timber; special forest products and forest botanical products", 18 January

Kimmerer, R. and F. Lake (2001) "Maintaining the mosaic: The role of indigenous burning in land management", *Journal of Forestry*, vol 99, no 11, pp36–41

Lightfoot, K. and O. Parrish (2009) *California Indians and their Environment*, University of California Press, Berkeley, CA

Lynch, K. and R. McLain (2003) *Access, Labor, and Wild Floral Greens Management in Western Washington's Forests*, USDA Forest Service Pacific Northwest Research Station, PNW-GTR-585, US

McLain, R. J. and E. T. Jones (2005) *Non-Timber Forest Products Management on National Forests in the United States*, USDA Forest Service Pacific Northwest Research Station, PNW-GTR-655, US

McLain, R. J., E. M. McFarlane, and S. J. Alexander (2005) *Commercial Morel Harvesters and Buyers in Western Montana: An Exploratory Study of the 2001 Harvesting Season*, USDA Forest Service Pacific Northwest Research Station, PNW-GTR-643, US

McLain, R. J., S. Alexander, and E. Jones (2008) *Incorporating Understanding of Informal Economic Activity in Natural Resource and Economic Development Policy*, USDA Forest Service Pacific Northwest Research Station, PNW-GTR-755, US

NAHC (California Native American Heritage Commission) (undated) "Letter in response to the January 29, 2009 Federal Register Rule Delay Notice of USDA Forest Service 26 CFR Parts 223 and 261"

News & Views (2000) "New legislation would regulate Forest Service's NTFP management", spring, *News & Views: Communities and Forests*

Nuvamsa, B. (2008) "Letter to forest management director, USDA Forest Service, regarding the proposed special forest products rule", 22 January 2008

Ortiz, B. (2008) "Re: Sale and disposal of national forest system timber; special forest products and forest botanical products", 36 CFR Part 223, 22 January

Parker, L. and J. Kalt (2008) "Letter to the director, Forest Management Staff, US Forest Service, in response to special forest products and forest botanical products (36 CFR Part 223)", 22 January, California Indian Basketweavers' Association

Prucha, F. P. (1994) *American Indian Treaties*, University of California Press, Berkeley, CA

Ramos, J. (2009) "Re: Final Rule Amending 36 CFR Parts 223 and 261: Sale and Disposal of National Forest System Timber; Special Forest Products and Forest Botanical Products" , 2 March, Tribal Chairman, San Manuel Band of Serrano Mission Indians

Robles, R. and A. Mooney D'Arcy (2009) Letter to director of forest management, USDA Forest Service, entitled "Objection to proposed system of permits and fees for traditional gathering – special forest products final rule", 27 February, Co-directors, United Coalition to Protect Panhe

Schlosser, W., K. Blatner, and R. Chapman (1991) "Economic and marketing implications of special forest products harvest in the coastal Pacific Northwest", *Western Journal of Applied Forestry*, vol 6, no 3, pp67–72

Solberg, D. (1999) "Uncommon bounty", *High Country News*, 15 February

Stone, V. (2009) "Comment letter concerning the proposed special forest products rule", 2 March

Thomas, M. G. and D. R. Schumann (1993) *Income Opportunities in Special Forest Products – Self-Help Suggestions for Rural Entrepreneurs*, Agricultural Information Bulletin AIB-666, USDA Forest Service, Washington, DC

United Nations (2007) *United Nations Declaration on the Rights of Indigenous Peoples*, UN General Assembly Resolution 61/295, adopted 13 September 2007, www.un.org/esa/socdev/unpfii/en/drip.html

Villarruel, S. (2009) "Re: Federal register issuing a notice of final rule on the sale and disposal of national forest timber; special forest products and forest botanical products", 12 February, Environmental technician/cultural information officer, Pit River Tribe

Walters, H. (2009) "Midnight reg runaround: Forest rule would make even granny basket weavers go get a permit", *North Coast Journal*, 5 February

Weingardt, B. and M. Pool (2006) "Memo to area managers and forest supervisors", Regional forester, Pacific Southwest Region, Forest Service, and California state director, Bureau of Land Management, 29 November

Part IV

What Authorities Recognize Forest People's Rights?

Claims on forests require sanctioning by authorities to be recognized as rights, as we have highlighted in Chapter 1. This is particularly apparent in situations when multiple actors assert competing claims on forests, just as those discussed in the preceding chapters. People's claims only turn into enforceable rights once they find support by an institution of authority. Such an institution can be the state or one of its constituent parts, such as national legislation. Yet, it can also be a customary leader, religious body or transnational convention which accords legitimacy to a claim and thereby recognizes it as a right. Moreover, claims only become rights if recognized through authoritative procedures (i.e. procedures employed by institutions of authority that are recognized as legitimate). A national government, for example, may be recognized as possessing authority over forests, although it remains bound to particular procedures for the recognition of rights.

The two chapters included in Part IV look at the relations between forest rights and authority. They consider various institutions considered to possess authority over forests, particularly the state. The emphasis on the state reflects a central concern with state recognition, which is shared widely among forest rights activists. The underlying rationale is that activists consider state recognition to provide the most robust forest rights in most circumstances. The chapters also look at the procedures through which politico-legal institutions recognize claims on forests as rights, contrasting the formal procedures established for the exercise of control with the actual workings of authority.

In Chapter 11, Larson and Cronkleton examine various cases of institutional competition about authority over forest rights. They show how forest people may be subject to further processes of negotiation over authority after having accomplished collective rights-recognition goals through grassroots-level endeavors; the appropriate authorities have to be selected and recognized in order for communities to act upon the rights allocated to them as collective entities. Drawing on insights from Nicaragua, Bolivia, the Philippines, and Ghana, Larson and Cronkleton demonstrate that the choice of these institutional representatives often has both material and symbolic importance. Hence, the process of identifying them may be subject to substantial dispute and conflict, particularly between communities, political leaders, and state entities. Issues of institutional selection are both complex and variable, as these different actors seek "representatives" that serve their own priorities and interests. These insights thereby demonstrate that after achieving collective rights recognition by the state, forest people encounter new challenges of representation, with significant effects on the distribution and sharing of land, resources, and political powers.

In Chapter 12, Dorondel takes a close look at the practices and procedures employed by local government in a forest village of Romania. In theory, he suggests, local government can play a key role in reconciling the conflict between local people's interests in logging and wider conservation interests in protecting Piatra Craiului National Park. Local government could help to recognize villagers' historical forest rights and enforce environmental safeguards, which would allow logging in restricted areas. Yet, in practice, the local mayor does not seek to mediate between the conflicting interests, but exploits the conflict for his personal advantage. He uses the designation of the park to influence the forest restitution process for his own personal gains. Operating a logging company together with his wife, he also employs the national park as a threat to buy timber from villagers cheaply. In this way, his practices compromise both villagers' forest rights and conservation goals, aggravating the conflict between villagers and the park officials. These insights, therefore, illustrate the commonly encountered discrepancy between formal and actual state practices and its implications for forest rights.

The two chapters thus demonstrate that effective rights recognition depends on supportive authority, as do many contributions to other parts of the book. In other words, rights only serve forest people if sanctioned by suitable institutions and practices of authority. Suitable institutions may include the state and its constituent parts, such as the national and local governments examined in the two chapters. In addition, rights recognition may come from customary arrangements (Chapter 11), local public–private partnerships (Chapter 9), transnational conventions (Chapter 3), religious institutions, or other institutions of authority. Suitable procedures, in turn, may include legislative action (Chapters 3 and 12), decision-making by representative collective bodies (Chapters 9 and 11), and administrative decisions made by government officials (Chapters 8), to name just a few of the more prominent ones.

The chapters also show that no institution or procedure is inherently authoritative. The authority attributed to an institution or procedure ultimately rests on the question if involved actors consider it to legitimately sanction claims as rights. Thus, we once again find that universal notions of authority run into problems. Just as the kinds of claims that people make on forests are not uniform (Part II), and just as it is impossible to universally define the sorts of actors to support (Part III), the recognition of rights requires context-specific approaches. The context specificity of authority applies to all kinds of settings, including those characterized by apparently strong and functioning state structures. Who, after all, would have expected to detect such blatant patronage relations in Romania, a country that acceded to the European Union in 2007?

These insights open up a new perspective on another longstanding debate amongst forest rights activists on the authority attributed to customary arrangements and traditional leaders. People in one camp have questioned the legitimacy of traditional authority, often referring to empirical insights from sub-Saharan Africa documenting traditional leaders' corrupt practices. People in the other camp have used other empirical evidence, often from Latin America and Southeast Asia, to assert the legitimacy of customary arrangements, particularly in comparison with local and national governments' lack of accountability to forest people. The insights presented here show that the primary question is not about the essential features of customary authority and its distinctiveness from governmental authority. The primary issue is the authority attributed to different institutions and procedures by various kinds of actors in particular contexts. Understood this way, it becomes clear that in some settings, rights recognition by national governments through customary land titling and other means may be the most effective strategy for securing forest people's rights. Yet, in other settings, forest people's rights may be served much better through endorsement of customary authority over forests.

The insights from Part IV also suggest that it may be useful for forest rights activists to think about nested forms of authority instead of focusing on single institutions and universal procedures, such as state recognition through legislative reform. Forest people's actual claims may receive the strongest sanctioning as rights under authority relations that bring together multiple institutions and combine various procedures. Nested authority may provide a place for transnational conventions, national governments, local governments, customary arrangements, village leaders, etc., and utilize legislative action, collective decision-making, administrative decisions, court resolution, public–private deliberations, etc. The specific institutions and procedures constituting nested authority will vary from place to place. In this way, nested institutions and procedures provide the possibilities for the context-specific and adjustable definitions of rights highlighted in Part II of this book. Nested authority also facilitates the context-specific support to concrete actors and use of inclusive rights notions underlined in Part III.

11

Who Represents the Collective? Authority and the Recognition of Forest Rights

Anne M. Larson and Peter Cronkleton

As mentioned previously in this book, a number of countries, particularly since the mid 1980s, have begun to recognize community rights to forests (White and Martin, 2002; Agrawal et al, 2008; Sunderlin et al, 2008; Larson et al, 2010a; see also Chapter 2 in this volume). These processes have often evolved in response to grassroots demands, such as those of indigenous communities, for the recognition of rights to forests that they have managed or used historically under customary institutions. Recognition may involve rights to resources or to resource revenues that were not previously acknowledged, and it may go so far as to involve demarcation and land titling.

What does *authority* have to do with the recognition of communal rights to forest (see also Chapter 1 in this volume)? The idea of "recognizing" rights implies a simple process of giving one's blessing, in this case the state's legal blessing, to something that already exists. The relevant definitions in Webster's dictionary define the term "to recognize" as "to admit the fact of" or "to acknowledge formally" (Webster, 1967). But the reality of recognizing people's rights to land or resources is a far more complex process. This chapter looks specifically at issues of authority as they become apparent in three different ways.

First, recognizing tenure rights involves choosing an entity or person to be the legal representative of the rights-holders when rights are granted (see

Fitzpatrick, 2005). Even in cases where the names of all the people receiving rights appear on a land title, an entity is usually needed to act on behalf of the group. The title or right is often granted to this representative.

Second, establishing this representative involves defining its domain of powers, or sphere of competence (Fay, 2008). What external negotiations can this entity enter into on behalf of the community? What power does it have over community members' access to resources (see also Chapter 5 in this volume)?

Third, defining collective rights-holders and their representative is intimately tied to the definition of the physical space – the land area and resources – to which rights are being recognized. On the one hand, the specific spatial configuration, as through the demarcation of borders, determines who has rights to the area in question and who does not, with obvious consequences. On the other hand, the definition of a territory may have broader implications, playing a central role in geopolitical negotiations (see Sikor and Lund, 2009), such as between indigenous peoples and the state (Larson, 2010).

Each of these issues constitutes a potential point of contention between the community demanding rights recognition and the state, or the specific state entity granting legal recognition. The chapter explores four cases, in Nicaragua, Bolivia, Ghana, and the Philippines, and demonstrates how the process of recognition may lead to conflict over authority, or to the construction of new, legitimate and effective representation.

Authority Relations and Communal Tenure

The term "authority" is used in several ways, particularly in the realms of policy and practice. In particular, it is used to refer both to the abstract notion of power (e.g. to hold authority) and to the person or institution holding that power (Fay, 2008). According to Weber (1968), authority refers to power that is "legitimate". The issue of legitimacy then raises the question: legitimate to whom? If authority requires legitimacy, it cannot be a fixed attribute that is mandated or assumed. Rather, it must be constructed through social interaction and is subject to conflict and negotiation (Sikor and Lund, 2009).

The central issue of concern in this chapter is the entity selected to represent the collective that receives formal rights under new legal arrangements (see Ribot et al, 2008). Both the nature of the institution representing the collective and its domain of powers are fundamental to the distribution of access to land and forest resources and to the benefits they generate.

When people receiving new or formal rights already have customary rights to the land, it might seem that the simplest solution is to recognize the entity that is currently in power. There are, however, at least two problems with this. First, formally recognizing an institution changes it: when the state recognizes a particular institution as the community representative, it is granting that institution external legitimacy (Ribot et al, 2008). But this institution may not have internal legitimacy, or it may have internal legitimacy, but not to manage the particular set of powers now being granted (Fay, 2008). An example is the recognition of non-democratic institutions – chiefs and headmen who inherit their posts – in some African nations undergoing decentralization (Ntsebeza, 2005; Ribot et al, 2008).

Second, the granting of tenure rights may necessarily involve the formation of new entities to represent beneficiaries, particularly for large territories. Indigenous movements in several Latin American countries, including Bolivia and Nicaragua, have promoted a territory model, comprising multiple communities, for the implementation of indigenous property rights. But this is a scale at which governance institutions do not currently exist. Hence, defining both the institution and the domain of powers involves forging new ground. The Nicaragua, Bolivia, and Philippine cases here all represent multi-community indigenous territories.

The issues mentioned so far present a somewhat simplified view of these political processes – that is, "choosing" or "forming" a new authority or "defining" a domain of powers is no easy task. Given the material as well as symbolic importance of the outcome of these processes, it is not particularly surprising that they would be subject to conflict.

The demands for recognition themselves have usually emerged from conflict, including the righting of historical wrongs, such as the treatment of indigenous people in Latin America. In cases involving land titling, demarcation is subject to conflict and negotiation over the definition of borders, as well as over the fate of "outsiders" holding land inside the territory. But the issue of control over territory may also be part of larger geopolitical struggles.

The Case Studies

This section presents four cases of forest tenure reform included in a study by the Center for International Forestry Research (CIFOR) from 2006 to 2008.[1] The Nicaragua and Bolivia cases refer to large indigenous territories being demarcated and titled in the wake of important changes in national legislation to recognize indigenous land rights. The Nicaraguan study highlights conflicts between indigenous leaders and communities over the configuration of territories and "territorial authorities" in the North Atlantic Autonomous Region (RAAN). The Bolivian case discusses the demarcation and titling of the Guarayos indigenous territory. In that case, the organization that led the battle for land rights also served as the representative of the Guarayos people, but was granted ambiguous authority and eventually split into two over corruption charges and power struggles between political factions in the regional and national government.

The Ghana case refers to recognition of the rights of forest-edge communities to participate in the distribution of royalties from the sale of forest products. Traditional authorities are the official recipients, and little reaches communities. The final case refers to another indigenous territory, the Ikalahan ancestral domain, in the Philippines. In contrast to the other cases, the official territorial authority, the Kalahan Educational Foundation (KEF), maintains a high level of legitimacy as a representative local authority.

Indigenous territories in the RAAN, Nicaragua

Indigenous peoples in Nicaragua won formal rights to their communal lands in the 1987 constitution, and during the same year, the Autonomy Law created the North and South Autonomous Regions. These regions represent about 45 percent of the national territory, but only 12 percent of the population (INEC, 2005). The first autonomous regional governments were elected in 1990.

After indigenous land rights were formally recognized in the constitution, it took another 15 years and an international court case (see Anaya and Grossman, 2002) for the National Assembly to pass the Communal Lands Law (Law 448). Until this time, the central government continued to treat the region's natural resources as state property. The regional governments were very weak and had little funding or power. Three things began to change this: the Communal Lands Law (in 2003), the long-awaited approval of the implementing regulations of the Autonomy Statute (also in 2003), and, most importantly, a change of government (in 2007). The first titles were delivered in late 2006; but it was thanks largely to an alliance between the entering government administration in 2007 and the Miskitu political party, Yatama, that most of the indigenous territorial claims in the RAAN had been titled by 2010.

The communal lands law formally recognizes indigenous land rights and also establishes the institutional framework for demarcation and titling. Communities have the right to choose how they want to be demarcated – as a community or multiple communities – and, in the latter case, to elect their territorial authority in an assembly of all the communal authorities (Articles 3, 4). This new governance institution is the territory administrator and legal representative (Article 5). The regional government then registers and certifies the people elected. According to the law, the elected "community authority" is in charge of community land and resources; the "territorial authority" is in charge of resources common to the multiple communities of a territory (Article 10).

Two groups of communities were studied in-depth: Tasba Raya and Layasiksa. The former had decided to form a seven-community territory in 2005; the latter expanded from a two- to a three-community territory in 2009. Both designed their territories based on common history and affiliation as a group of communities, and both had elected their territorial authorities according to the procedures established in the law. Nevertheless, the autonomous government would not provide accreditation, and indigenous political leaders refused to recognize their territories.

Political leaders from Yatama pressured communities to form territories based on a design of their own conception. According to Miskitu leaders, they were interested in forming territories that covered a significant part of the land area, and moving quickly while the political moment was favorable in order to position themselves "in between" the central government and the region's communities and resources (CRAAN, 2007). More importantly, their design involves reshaping the region's electoral districts; the municipal structure imposed by the central government would be eliminated and replaced with an "indigenous" structure of territories and territorial authorities. This cannot happen officially, however, without legislative reforms.

In theory, if community self-government was the foundation, with multi-community territorial institutions at the second tier, comprising electoral districts for the election of regional autonomous councils, this new governance structure could provide the institutional basis for the self-determination of the indigenous and ethnic populations of the autonomous regions. But not all indigenous and ethnic groups, even many Miskitu, feel represented by Yatama or trust its leaders' motivations. In the two territories studied, the lack of accreditation of their elected authorities had already had concrete consequences, including the communities' inability to access funds designated for the territory. Both territories were subsumed into larger territories according to Yatama's design. And though the law mandates that territorial authorities be elected, it appeared, at least in some cases, that they were being handpicked by political leaders.

The Nicaraguan case demonstrates how property borders became the negotiating ground in a larger battle between indigenous leaders and the state for legitimate power over the region. This issue has shaped the nature of representation – that is, the member communities should elect the legal representative of their territory; but, in practice, regional government authorities have refused to recognize these leaders, in part because they have refused to recognize the territories that were being contemplated or designed by communities. Indigenous leaders have used territorial strategies to consolidate their power *vis-à-vis* the central government in a broader, legitimate geopolitical struggle: central government administrations have tried to control the region's resources, and this broader plan positions "the region" better for the future. Hence, the political configuration of territory is deeply linked to the struggle for economic power and control over natural resources.

Nevertheless, the process has sidelined the needs and desires of the communities whose rights were being recognized, and placed indigenous leaders at odds with indigenous communities, at least in some cases. These leaders have sought to impose their own territorial "representatives", whose domain of powers includes allocating natural resources, both internally and with external actors, that are common to those communities. In other words, defining the configuration of territories became a way to strengthen indigenous political power *and* to control community resources.

Guarayos TCO, Bolivia

The recognition of indigenous land rights in Bolivia has resulted from a slow process of policy reform driven by mass marches and other protests to pressure government decision-makers. One result was the recognition of a new type of indigenous property, known as original community land (*tierra comunitaria de origen*, or TCO), in the agrarian law of 1996. TCOs are communal properties that are granted based on ancestral claims combined with needs assessments carried out by the government.

The Guarayos' TCO is located in a rapidly changing forest frontier province of the same name in the north of Bolivia's Santa Cruz Department. The land claim emerged, in part, due to tensions arising from competition for land after an interdepartmental highway led to an influx of logging industries, ranchers, large-scale commercial farmers, and smallholder colonists. Hence, in 1992, the Guarayos people created the Central Organization of Native Guarayos Peoples (COPNAG) to pressure for their claim. Its leaders were elected by traditional community-level organizations in 12 indigenous villages and collective neighborhood organizations in the 6 larger towns representing the Guarayos population across the province. In 1996, COPNAG presented a TCO demand for almost 2.2 million hectares, which was reduced to 1.3 million hectares after the government's spatial needs study (VAIPO, 1999). Through a rule referred to as "immobilization", new third-party claims in the area were prohibited until titling is completed. Nevertheless, in one important move that undercut grassroots confidence in the process, the government renewed logging concessions to more than 500,000ha of forests in Guarayos, much of which was inside the territory, in spite of COPNAG's protest that these constituted "new claims" (Vallejos, 1998).

Demarcation involves the evaluation and "regularization" of third-party claims before issuing collective titles, a complicated process given that non-indigenous peoples make up a significant percentage of the population in some areas of the territory. Legitimate claimants include people with long histories in the region or those already with title. In this process, COPNAG was given the role of certifying the authenticity of pre-existing land claims by non-indigenous peoples. In fact, COPNAG went from being an advocacy organization to a land administrator and official representative of Guarayo interests to the government. It would hold the TCO land title in the name of the Guarayos people and played a role in forest resource allocation, as well. However, the power granted is ambiguous because the Guarayos TCO is superimposed over three municipalities that have official mandates and budgets, while COPNAG does not.

Demarcation of the TCO moved quickly at first because the tenure reform agency chose to start in remote areas. Hence, by the end of 2003, about 1 million hectares had been titled. But by late 2006, only an additional 18,000ha had been titled and little progress had been made near the highway and main town, where most of the population is concentrated and property rights are most contested. Among other things, long delays and the strategy to avoid conflictive areas in the early stages allowed illicit land transactions to take place in the accessible lands that were highly prized by both indigenous peoples and outsiders. Competing claims often involved economically and politically powerful individuals, and COPNAG leaders were implicated in providing forged certification documents for landowners (López, 2004; Moreno, 2006). Charges surfaced that there had been 44 fraudulent transactions in 2001, involving private landowners, COPNAG leaders, and technicians from the tenure reform agency (López, 2004).

The accusations of fraud and the influence of competing interests generated turmoil in the Guarayos political movement and led, in 2007, to the expulsion of tainted leaders and the election of COPNAG's first female president. Parallel to the internal Guarayos conflicts, Bolivia was beset by increasing tension due to a power struggle between the departmental and national government over, among other issues, regional autonomy. Regional leaders associated with the *Comité Cívico* of Santa Cruz sought to sow division among President Morales's indigenous base. They encouraged the expelled Guarayo leaders to form a parallel group they call the "authentic" COPNAG, which the departmental government then recognized as the legitimate representative of the Guarayos TCO. As a result, the Guarayo organization is divided much along the contours of the national political conflict between the central government (in favor of the indigenous president, Evo Morales) and regional departmental governments (against Morales and demanding regional autonomy).

In Guarayos, communities elected their indigenous territorial representative, though COPNAG was originally an advocacy organization that was given a new domain of powers under changing circumstances. At the same time, its influence on the titling process was undermined by decisions such as the one to renew logging concessions, and its domain of powers beyond the titling process were ambiguous in relation to existing municipal governments. Political and economic pressures, a new domain of unfamiliar and ambiguous powers, and lack of oversight and accountability resulted in corruption and a split in the leadership. As in Nicaragua, the process of representation was fundamentally shaped by the broader geopolitical struggle for indigenous rights *vis-à-vis* central government. When the process began, vested state interests interfered with titling at both central and regional levels. More recently, conflict between departmental and central governments increased, particularly after the Morales administration placed central government support behind the titling of TCOs.

Distribution of timber royalties, Ghana

Although the state legally manages Ghana's natural resources "for the benefit of the population", important tenure reforms have occurred since the return to constitutional rule in 1992. The constitution and several other laws now directly promote the sharing of benefits from forest products with forest-edge communities. For example, the formula for the distribution of stumpage fees, established by the constitution, mandates 25 percent to the "stool" (a family or clan represented by a chief or head of family), 20 percent to the traditional authority (presumably the paramount chief), and 55 percent to the local government. This discussion will focus on the stumpage fees distributed to the first two entities.

Ghana has elected district assemblies; but land and resources are tied to the traditional landownership system and the institution of chieftaincy – both are preserved in the 1992 Constitution (Article 267). And while natural resource control and management are vested in the president, traditional authorities are the owners. The chieftaincy position is hereditary, based on membership in a royal family or clan belonging to a community that has collective ownership of a specific portion of land, called stool land. A traditional area is an area within which a paramount chief exercises jurisdiction – that is, traditional areas are not linked with state administrative boundaries, but rather associated with a paramount chief. Each paramount chief presides over two ranks of sub-chiefs, as well as the elders in a traditional area's council. Customarily, all the land in a traditional area is "symbolically" under the paramount stool; but ownership is complex and subject to multiple claims, especially by the lower-level chiefs. Chiefs are important and powerful leaders in Ghana because they have a certain legitimate claim of custodianship over community properties, and rule over specific territories and domains. In short, their jurisdiction has permitted them to assume positions as community representatives since colonial times.

As stated above, forest revenue is supposed to be distributed to the stool and to the traditional authority. The law is ambiguous, however, stating that funds should be directed "to the stool through the traditional authority for the maintenance of the stool in keeping with its status" (Article 267). There is no explicit requirement that the stool's 25 percent be reinvested in the community. Nor is it clear whether local people should benefit from the other 20 percent allocation to traditional leaders in their private capacity.

According to Opoku (2006):

> *Chiefs tend to appropriate royalties for their personal or household use and have often claimed that this is the meaning of "maintenance of the stool in keeping with its status"; yet the Constitution clearly states that land ownership carries "a social obligation to serve the larger community", and that stool managers are fiduciaries "charged with the obligation to discharge their functions for the benefit ... of the stool".* (Article 37, Section 8)

The allocation of forest revenue is also contested based on multiple claims within the hierarchy of traditional authorities. Some subordinate stools complain that payment of royalties through the traditional authority enables paramount stools to appropriate part or all of these funds (Opoku, 2006). Marfo (2006) documented one such conflict between the chief of the Juaso stool lands and the Dwaben paramount chief. The former argued that he and his elders constituted the traditional authority over the Juaso stool land and thus should receive the Juaso stool revenue directly. The paramount chief argued that he owned all lands in the traditional area over which he exercised jurisdiction and should thus receive the stool revenues.

The Ghana case does not involve land titling, but rather the recognition of rights to royalties from the sale of forest products, which are managed by the state. Although different from the other cases, the problem of legitimate representation is similar. In this case, chiefs are mandated by law to be the recipients of funds allocated to the local arena. They may have legitimacy as leaders or even representatives of communities in other spheres, but the receipt of logging royalties constitutes a new domain of powers. This case demonstrates the problem with recognizing existing traditional authorities as community representatives without questioning or guaranteeing this representation: there were conflicts between different levels of chiefs, royalties were not reaching communities, and chiefs justified their personal use of resources due to ambiguous language in the law.

KEF, the Philippines

The first indigenous community in the Philippines to receive recognition of its forest rights was the community of the Ikalahan people, which obtained a 25-year agreement for the right to use, manage, and exclude third parties from the Kalahan Forest Reserve in 1974. They finally received ownership rights – a permanent certificate of ancestral domain – in 2006. The Kalahan Educational Foundation (KEF) (originally set up to establish a high school, hence the name) is the formal representative of the tribe and the designated institution with decision-making power over land and forest management.

The struggle of the Ikalahan people for the formal recognition of their rights began during the late 1960s. In the wake of a successful court battle in 1972 against encroachment from land-grabbers, the Ikalahan decided to form an organization to fight for formal recognition of their claim. With the assistance of an American missionary, Pastor Delbert Rice, they formed the KEF.

The KEF has about 500 member households in seven communities (*barangays*, the smallest units of political administration). More than 90 percent of the people living in the reserve are Ikalahan, and all Ikalahans are automatically KEF members. In each village, the adults in each *barangay* constitute the Barangay Assembly and are all voting members. Each *barangay* has elected local government officials (the *barangay* council), tribal elders (almost always men), and informal tribal leaders. According to Rice (2001), elders hold office by ascription and are recognized as being effective at providing leadership and resolving disputes; but they do not represent the community or make decisions for the community. The most important institution is the Tongtongan. The Tongtongan functions like a tribal court, presided over by local elders, whereby the community comes together to discuss a conflict or problem; the elders make the final judgment, which is aimed at reconciliation (Rice, 1994). The Tongtongan, as an informal or customary institution, is even more important for decision-making than the KEF.

The KEF was formed by a group of elders, and its first board of trustees was made up of one representative from each of the participating *barangays*, plus three others. Its configuration has evolved over time, and today there are 15 voting members, including 3 women. The *barangays* each choose their representatives for two-year terms in general assembly meetings, which are held twice a year. The KEF is charged with establishing and enforcing the rules for the reserve. Today, these include regulations regarding swidden farming, tree cutting, chainsaw registration, fishing, quarrying, hunting, and land claims. It approves the allocation of all household parcels, as well as land transfers among tribal members.

The relationship between KEF and *barangay* governments is based on trust and mutual cooperation, including shared revenue from timber permits (Dahal and Adhikari, 2008). Community members also largely respect the rules, which were presented and discussed in each *barangay* before final approval by the board of trustees. The regular general assembly meetings are open to all, and when important issues need to be discussed, attendance and participation are high (Dizon et al, 2008). The Tongtongan works hand in hand with the KEF governance system. Honesty, equity, and fairness are explicitly promoted. In one case, the chair of the board was implicated in illegal harvesting and transport of timber from the forest, and he was penalized (Dahal and Adhikari, 2008). A third-party financial audit is conducted every year. Pastor Rice, who played an important role in building social capital and encouraging fair internal management, serves as executive director of the KEF and helps mediate relationships between the community and external actors, such as the government, donor agencies and non-governmental organizations (NGOs).

As in Guarayos, the organization chosen to represent the Ikalahan collective in the Philippines case was the same one that was originally set up to fight for land rights. But the KEF is a success story. It had a number of advantages over COPNAG: the territory is much smaller and more homogeneous, facilitating communication for accountability; powerful outsiders seeking to invade Ikalahan lands were defeated in court over 30 years ago and the current situation is much more conflictive in Guarayos. Also key to the KEF's success, however, has been finding the appropriate balance between the new entity with powers over land and natural resources and traditional institutions, together with significant efforts to guarantee transparency and accountability. It is notable that a trusted, embedded external broker has facilitated these processes.

Lessons on Authority in the Recognition of Rights

The cases present different ways in which authority relations have played out in four distinct contexts of rights recognition. Together they provide an instructive panorama of the issues surrounding authority that emerge in the recognition of communal land rights. Although many of the specifics vary, the common thread is the central importance of the entity that has been chosen to represent the collective – and the complex dynamics surrounding the establishment of this "authority". This section first revisits the three issues regarding authority discussed in the introduction, and then summarizes a set of key lessons emerging from the cases.

The legal representative. In the three indigenous territories, new entities to represent rights-holders had to be created, as governance institutions did not previously exist at the territorial scale. In theory, all of these entities are elected by the communities that are members of the territory, but only in the Philippines' case has this process run smoothly. In both Nicaragua and Bolivia, political leaders and communities sought *different entities* as the legitimate territorial representative. In both cases these entities were only one aspect of a much larger conflict, demonstrating how the battles over territory and representation are inextricably linked. For its part, the Ghana case highlights the difficulties of simply choosing an existing traditional leader to "represent" a community. Not only did funds fail to reach community members, but also different traditional leaders vied for the right to those funds.

Domain of powers. In all four cases, the entity selected to represent communities has an important domain of powers. In the indigenous territories, these leaders have power over people and resources inside the territory, and also serve as the legal representatives of community interests externally. Hence, communities depend on them, at least in part, for access to land and resources, and for political representation of their interest. In Ghana, chiefs have also been granted power over resources – in this case, a portion of funds, presumably intended for investment in the community, that have been raised from the sale of the community's forest resources. The powers over collective resources that each of these entities hold make them a target for external actors seeking control over people and resources on the ground.

Defining territory as geopolitics. The recognition of rights to territories in the three indigenous cases emerged from conflict. The demand for rights was often precipitated by direct incursions into lands claimed by indigenous communities, but it is rooted in broader-scale struggles relating to the denial of indigenous rights, historically. It is no surprise, then, that the recognition of indigenous rights may not signify an end to conflict, but rather the beginning of a new phase of struggle. This is particularly clear in the two Latin American cases, which involved much larger expanses of territory, as well as demands for autonomy and self-governance. These two cases demonstrated that struggles over territory and authority are intimately linked. Territory leaders, or authorities, become key loci around which issues of political and economic power converge.

Six lessons emerge from the cases. First, the entities chosen to represent communities or territories matter. These authorities – legitimated by the state through the recognition process, and by communities through election, etc. – have concrete effects on outcomes for communities. In each case, they allocate natural resources and/or have access to funding for the community.

Second, apparently simple solutions, such as recognizing the existing "authority", may not, in fact, be simple at all, leading to conflict and failing to produce the intended result of increasing rights for the entire community. The Ghana case demonstrates how this option resulted in conflict among different levels of chiefs, regarding which level should receive the designated funds, but notably little debate – among chiefs, at least – regarding how those funds should benefit communities.

Third, even when communities elect their representative and defend the local legitimacy of this authority, a state entity (or other key actor) may have conflicting interests. This is what happened in the RAAN. Hence, representation at the territory level, tied to the configuration of territories, became the battleground with indigenous leaders, who in turn sought to reshape the design of representation at the regional level to their political advantage, *vis-à-vis* the central government.

Fourth, the election of entities at the territory level may lead to overlapping, ambiguous, and conflicting domains with existing state government structures, such as municipalities. In Guarayos, for example, the indigenous territory crosses municipal borders, and their specific relationship and distinct domains of powers have not been defined. In Nicaragua, if indigenous leaders are unable to get legislation passed to replace the existing municipal structures with territorial boundaries, the problem of jurisdiction will also have to be resolved.

Fifth, elected, representative, and locally legitimate authorities can break down and lose legitimacy, particularly under the weight of a new and unfamiliar domain of powers in unfavorable circumstances. Such circumstances surrounded the COPNAG leadership in Guarayos: serious political opposition, even within key government implementing agencies; heavy pressure from powerful economic interests and individuals; and a difficult territory that was large and discontinuous, ethnically diverse, impoverished, and with little infrastructure over large areas.

Sixth, effective representation is possible. The KEF is an effective organization with high levels of both internal and external legitimacy. It is neither entirely new nor entirely traditional, but has found, over time, an acceptable balance between the two. A highly respected, embedded external broker played a key facilitating role. This does not mean that an external actor is a necessary condition for the construction of authority, but the role this actor played may be – that is, effective representation requires transparent rules of the game, including broad agreement on how representatives are chosen and their specific domain of powers, and the incorporation or design of appropriate accountability mechanisms. This is also a learning process that evolves over time.

Conclusions

This chapter has demonstrated that the issue of authority should not be ignored or treated lightly in the process of recognizing rights to forest or land. The recognition of rights is often contentious and is likely to result from grassroots struggle – and there is no reason to believe the struggle ends once rights are granted (see Larson et al, 2010a). One key arena of contention is the choice of entity to represent the collective, an issue intimately tied to the control of land, resources, and political power. Hence, it is no surprise that the "choice of authority" is subject to conflict and negotiation.

The four cases show that simply choosing the correct, downwardly accountable institution to represent those receiving rights may not be an option; in fact, in none of the cases did such an entity exist at the scale required. More to the point, legitimate power cannot be chosen: it has to be constructed.

Note

1 These case studies were adapted from Larson et al (2010b); the Ghana case was originally written by Emmanuel Marfo and has been adapted here, and Juan M. Pulhin provided valuable input on the Philippines case. For more on the cases, see Larson (2010) (Nicaragua), Cronkleton et al (2009) (Bolivia), Marfo (2009) (Ghana) and Dizon et al (2008) (the Philippines).

References

Agrawal, A., A. Chhatre, and R. D. Hardin (2008) "Changing governance of the world's forests", *Science*, vol 320, no 5882, pp1460–1462

Anaya, S. J. and C. Grossman (2002) "The case of Awas Tingni v. Nicaragua: A new step in the international law of indigenous peoples", *Arizona Journal of International and Comparative Law*, vol 19, no 1, pp1–15

CRAAN (2007) *Ayuda memoria: Asamblea territorial de Tasba Raya, Waspam, Llanos y Río Abajo y el Territorio MISRAT*, Municipio de Waspam Río Coco, Consejo de la Región Autónoma Atlántico Norte, Bilwi, Nicaragua, 19 May

Cronkleton, P., P. Pacheco, R. Ibarguen, and M. Albornoz (2009) *Reformas en la tenencia de la tierra y los bosques: La gestión comunal en las tierras bajas de Bolivia*, CIFOR and CEDLA, La Paz, Bolivia

Dahal, G. R. and K. P. Adhikari (2008) Bridging, Linking and Bonding Social Capital in Collective Action, Working Paper 79, CAPRi, Washington, DC

Dizon, J. T., J. M. Pulhin, and R. V. O. Cruz (2008) *Improving Equity and Livelihoods in Community Forestry: The Case of the Kalahan Educational Foundation in Imugan, Sta. Fe, Nueva Vizcaya, Philippines*, Project report, CIFOR and RRI, Bogor, Indonesia

Fay, D. (2008) "'Traditional authorities' and authority over land in South Africa", Paper presented at the Conference of the International Association for the Study of the Commons (IASC), 14–18 July, Cheltenham, UK

Fitzpatrick, D. (2005) "'Best practice' options for the legal recognition of customary tenure", *Development and Change*, vol 36, no 3, pp449–475

INEC (2005) *Resumen Censal: VII Censo de Población y IV de Vivienda*, www.inec.gob.ni/censos2005/ResumenCensal/Resumen2.pdf, accessed 10 April 2008

Larson, A. (2010) "Making the 'rules of the game': Constituting territory and authority in Nicaragua's indigenous communities", *Land Use Policy*, vol 27, pp1143–1152

Larson, A., D. Barry, G. R. Dahal, and C. J. P. Colfer (eds) (2010a) *Forests for People: Community Rights and Forest Tenure Reform*, Earthscan, London
Larson, A., E. Marfo, P. Cronkleton, and J. Pulhin (2010b) "Authority relations under new forest tenure arrangements", in A. M. Larson, D. Barry, G. R. Dahal, and C. J. P. Colfer (eds) *Forests for People: Community Rights and Forest Tenure Reform*, Earthscan, London, pp93–115
López, G. R. (2004) "Negociaron tierras fiscales en la TCO de Guarayos", *El Deber*, Santa Cruz de la Sierra, Bolivia, 7 November
Marfo, E. (2006) *Powerful Relations: The Role of Actor-Empowerment in the Management of Natural Resource Conflicts. A Case of Forest Conflicts in Ghana*, PhD thesis, Wageningen University, The Netherlands
Marfo, E. (2009) *Security of Tenure Reforms and Community Benefits under Collaborative Forest Management Arrangements in Ghana: A Country Report*, CIFOR and RRI, Accra, Ghana
Moreno, R. D. (2006) "COPNAG denuncia la venta en $US 1.2 millones de TCO en Guarayos", *El Deber*, Santa Cruz de la Sierra, Bolivia, 27 November
Ntsebeza, L. (2005) "Democratic decentralization and traditional authority: Dilemmas of land administration in rural Africa", in J. C. Ribot and A. M. Larson (eds) *Democratic Decentralization through a Natural Resource Lens*, Routledge, London, pp71–89
Opoku, K. (2006) *Forest Governance in Ghana: An NGO Perspective. A Report Produced for FERN*, Forest Watch, Ghana
Ribot, J. C., A. Chhatre, and T. Lankina (2008) "Introduction: Institutional choice and recognition in the formation and consolidation of local democracy", *Conservation and Society*, vol 6, no 1, pp1–11
Rice, D. (1994) "Clearing our own Ikalahan path", in J. B. Raintree and H. A. Francisco (eds) *Marketing of Multipurpose Tree Products in Asia*, Conference Proceedings of Multipurpose Tree Species Research Network in Asia, www.fao.org/docrep/x0271e/x0271e03.htm, accessed 10 January 2009
Rice, D. (2001) *Forest Management by a Forest Community: The Experience of the Ikalahan*, Kalahan Educational Foundation, the Philippines
Sikor, T. and C. Lund (2009) "Access and property: A question of power and authority", *Development and Change*, vol 40, no 1, pp1–22
Sunderlin, W., J. Hatcher, and M. Liddle (2008) *From Exclusion to Ownership? Challenges and Opportunities in Advancing Forest Tenure Reform*, Rights and Resource Initiative, Washington, DC
VAIPO (1999) *Identificación de Necesidades Espaciales TCO Guaraya*, VAIPO, La Paz, Bolivia
Vallejos, C. (1998) "Ascensión de Guarayos: Indígenas y madereros", in P. Pacheco and D. Kaimowitz (eds) *Municipios y Gestión Forestal en el Trópico Boliviano*, La Paz, Bolivia
Weber, M. (1968) *Economy and Society: Outline of an Interpretive Sociology*, University of California Press, Berkeley, CA
Webster (1967) *Webster's Seventh New Collegiate Dictionary*, G&C Merriam Company, Springfield, MA
White, A. and A. Martin (2002) *Who Owns the World's Forests?*, Forest Trends, Washington, DC

12

Tenure Rights, Environmental Interests, and the Politics of Local Government in Romania

Stefan Dorondel

Environmental safeguards are commonly proposed as a measure to reconcile local rights to forests with wider interests in environmental protection (e.g. Sayer et al, 2008). The underlying idea is a simple one: the transfer of formal property rights to local people, which is critical for recognizing local claims on forests, can be made compatible with environmental protection by enforcing certain restrictions on property rights. In this way, the logic is that local people receive property rights, but are also obliged to protect key environmental benefits. The use of environmental safeguards thus suggests a way out of the dilemma created by calls for the devolution of rights to local people: that local management may not consider the wider benefits provided by forests, such as the provision of downstream hydrological services or conservation of biodiversity habitats.

This chapter explores the potentials of environmental safeguards to reconcile conflicts between local rights and environmental protection. It discusses a case from Romania in which forests were given back to villagers at the same time as a national park was established. The chapter shows that conflict arose from these two simultaneous acts due to radical differences in the claims made by villagers and the park administration, and in the justifications underlying land reform and park establishment. The chapter furthermore indicates how the conflict was exploited by the local political elite to derive personal gains.

These insights suggest that arguments for the use of environmental safeguards need to be complemented with attention to the nature of the conflict between local rights and environmental protection, and to the functioning of the state and local power relations. In this case, the problem was not that the national park administration intended to preserve forest and biodiversity at a time when forest was given back to historical owners. The problem was that villagers' and park officials' claims were diametrically opposed to each other, and that state officials were accountable neither to the local population nor the wider public, but instead pursued their own personal gains.

The analysis performed in this chapter departs from two key theoretical propositions. First, one needs to examine the claims on the forest made by involved actors and their associated justifications to understand the nature of conflicts between local rights and environmental protection. Such analysis needs to answers questions such as: how do actors, ranging from private claimants to the state, build their property rights' claims? What methods do these actors use to legitimize their claims? In other words, one needs to analyze how different actors establish claims on forest, including attention to competing notions of rights to the forest and the quest to legitimize claims on forestland. Second, in order to understand the outcomes of such conflicts, one needs to take a close look at the state and local power relations. Powerful local people may take advantage of such situations to pursue their own advantages. Yet, depending on their accountability to the local population and the wider public, they may also seek to reconcile the conflict in different ways.

The chapter proceeds as follows: first, it examines changes in property relations during land reform in post-socialist Romania. Second, it turns to a case study to explore the restitution of forests to their historical owners: a process which ran parallel to the establishment of the national park. Third, it analyzes the ways in which the local political elite took advantage of the conflict between local claims and environmental protection – and how they profited economically from this. The chapter concludes with implications for the use of environmental safeguards to reconcile conflicts between local rights and environmental protection.

Property Relations and Land Reform in Post-Socialist Romania

During the early 1990s, the major concern of most post-socialist governments was to regulate property rights in their countries and do justice to those owners who were dispossessed in the period following World War II. Different countries have adopted different solutions to land and forest privatization. Romania opted for a *mélange* of two different land reform policies, offering both pre-1945 owners a form of historical justice through *restitution* of land and forest, and by *distributing* land to those people not entitled to restitution (Swinnen, 1997). As a result, land and forest restitution has triggered a "war of memory" among relatives, a competition among kin to find and prove their inheritance through documents (Verdery, 1998). Suddenly, cemeteries became a popular place to identify a dead uncle, an aunt, or a grandfather who had been long forgotten, as well as all potential ancestors who might have owned land and forest in the past. Thus, a competition of conflicting narratives was unleashed. People remember different stories about who was related to whom and revived villages' oral histories, which linked certain families to the foundation of the village (Dorondel, 2005). Land and forest restitution has not only initiated a quest to find a wealthy ancestor, but has also proven once more that property relations are linked to family history, and personal and community identity. This quest has also become an important part of post-socialist land relations (Hann, 2002; Hann and the Property Relations Group, 2003).

Land or forest restitution demands a long, complicated process for claimants. This includes writing an application (*cerere*), finding and producing documents proving their claims, and handing these in at the mayor's office. The request has to first go through the Local Land Commission (LLC), which includes the mayor, the secretary of the mayoralty (which in Romania is an official position), an agricultural engineer, the local representative of the National Department of Forestry (NDF), and a few elderly people in the village who supposedly know every family and the history of each piece of land. After the LLC approves the request, it is sent to the Regional Land Commission (RLC) for approval. Once ownership is established, the local and regional legal political institutions representing the state will then, in theory, enforce the rights and obligations of the owner. I say "in theory" because the post-socialist state, in many cases, has proven ineffective in enforcing state or private ownership of land and forest. In fact, land reform occurred simultaneously with the decentralization of the state and the shift of administrative and political decisions from the state to the local authorities. This fact has put local state officials (such as mayors, members of the local councils, or members of the local state offices, such as forest rangers or policemen) in an excellent position to take economic advantage of the implementation of land reforms (Mungiu-Pippidi and Althabe, 2002; Verdery, 2002; Dorondel, 2009). Consequently, acquiring statutory property rights of land and forest through land reform is different from exercising them in practice. "Effective ownership", as Verdery (2003) puts it, is more difficult to acquire for a range of reasons, including the local elites' interest in monopolizing valuable resources, the weakness of state agencies which enforce property laws, and political and social inequalities among villagers.

Forests proved to be the most valuable and marketable natural resource, and thus quickly became the most requested item in rural areas. The privatization of chemical and industrial factories during the mid 1990s led to an increase in agricultural input prices. The opening of the Romanian market in the early 1990s to highly subsidized Western produce and the drop in world market prices of agricultural products led to a crash in the price of agricultural products produced in Romania (Sarris and Gavrilescu, 1997). Thus, forests became far more interesting than agricultural land. Agricultural land was no longer a valuable natural resource, and it therefore did not tempt state or local officials. Conversely, the increasing demand for timber in the world market and the liberalization of timber prices in Romania led to an increase in the value of forests. As forests are a valuable natural resource, the state maintains as many rights as possible to co-manage privatized forestland. Besides, the forest, whether in public or private hands, fulfills environmental needs that motivated the state to become more actively involved in forest management than in land management. Thus, at least three groups of actors are interested in forests: state officials, the new forest owners, and the local elite. Each group has its own agenda, its own economic interests, and its own way of legitimizing its claims.

In many places worldwide, social actors make competing claims over forest as Moeliono and Limberg point out in Chapter 8 of this book. What makes the post-socialist case different is the double role played by the state in land reform. On the one hand, different state agencies, including the government, parliament, and the National Department of Forestry (NDF), are the legal-political institutions that are supposed to enforce the country's laws and rules. Laws are implemented through their local offices (such as the prefecture or mayor's office) and are enforced by legitimate state institutions (such as the police, forest guards, or park rangers). On the other hand, the state, as an owner, is also a major player. As any other owner would do, the state competes with other actors to acquire new property rights and to consolidate its ownership. The state prefers this ambiguous position. Not only must citizens ask state institutions to enforce their claims, but the state itself has also been interested in collecting property claims in order to reinforce them (Sikor and Lund, 2009). For the state, reinforcing property claims is a way of asserting its own power. This observation is particularly important in forestry.

Dominant groups at local and central levels, such as politicians, forest officials, and high-ranking policemen, use their power to access the more valuable natural resources (Hart, 1989; Stahl, 2010). While the state (in its ideal form) is meant to protect and enforce property rights, the local, rural elite are able to "hijack" state laws and regulations to fit their own interests (Verdery, 2002). During a period of major changes in agrarian relations, such as the transfer of property rights from the state or co-operative to private hands, access to government bodies represents an excellent opportunity for the rural elite to gain property rights themselves.

In conclusion, property is not only about rights and obligations, nor only about access to natural resources (Ribot and Peluso, 2003). Property also represents a complicated and continuous negotiation at different levels among different actors (Stahl et al, 2009). This chapter further explores the way in which actors legitimize their claims to the forest and shows the way in which the local elite use claims disputes for their own benefits. The following section begins by examining the intersecting history of the park and land reform in the village where I conducted fieldwork in 2004.

The Commune, the Park, and Forest Restitution

Dragova Commune

Dragova[1] is a mountainous commune in Arges County composed of three villages: Dragova (the administrative center of the commune), Podu Dambovitei, and Ciocanas. The commune's total population consists of 1100 people. A low valley surrounded by the Southern Carpathian Mountains, the commune resembles a bucket. The Bucegi Mountains lie in the east, and the Iezer-Papusa and Piatra Craiului peaks are visible in the west. Almost half of the commune's territory is part of the Piatra Craiului National Park (PCNP). A small creek divides the commune from north to south and constitutes the limits of the park. Although the area was not collectivized during socialism, Dragova, like any other commune in post-socialist Romania, experienced land reform. Following the collapse of communism, there were two major land reform laws: Law 18/1991, which stipulated that a maximum of 1ha of forest should be restituted to its former owners, regardless of how much forest the owner had in 1948; and Law 1/2000, which amended Law 18 by restituting up to 10ha of forest and also targeted the collective and communal forests. The villagers of Dragova obtained forest through three types of property rights: private, collective (called *Obşte*), and communal. A total of 132 families in the commune whose forest had been nationalized by the socialist government in 1948 received 280ha of private forestland. Most villagers actually owned more forest than they could prove. After 1945, when the communist regime came to power, most villagers were afraid that they would be declared *kulaks* (*chiaburi*), and therefore did not claim the actual amount of forest they owned.[2] In 1991, most people requested the forest that they or their fathers had owned; but if there were no documents available to prove their claim, they were denied ownership. The villagers see the state's refusal to consider their requests as theft.

Piatra Craiului National Park

The restitution of forest occurred at the same time as the foundation of the PCNP. The first law which established the reservation as a national park was issued in 1990. A total of 8100ha of forestland were also declared part of the national park, going much beyond the 440ha designated as a nature reservation in 1938. Between 1990 and 2004, the park was extended to 14,773ha. Of the 14,773ha, 6967ha lie in Arges County. The initial aim of the park was to protect the forest's biodiversity and to "maintain traditional land use" in this area (Plan, 2004, p2). Within the park, there is an amalgam of property rights. The state, individuals, and communities all have property rights to various pastures and forests within the park's borders. Individuals own 13 percent of the park's forest and almost 20 percent of the park's pastures. The state owns 40 percent of the forest and no pasture at all. While the state owns 42 percent of the park's territory, communities and private owners own the rest (Plan, 2004, p6). The state, through the *Regia Naţională a Pădurilor ROMSILVA* (NDF), has administrative rights to over 58 percent of the total territory of the park (including forest, meadows, and rock formations), while private owners, the *Obstea*, and six mayors' offices (of the localities next to the park) have administrative rights to the rest of the park's territory (Verghelet et al, 2003; Plan, 2004).

The forest in the park is zoned under several categories: the first is a special conservation zone of a total of 4476ha. The scientific reservation, where any activities besides scientific endeavors are forbidden, represents 683ha. A second zone is the "central area of the park" and its use, following IUCN Category II, is restricted to traditional activities such as grazing. Finally, there is a zone of protected land where, according to IUCN Category V, some economic activities such as rural tourism and forest use are allowed (Verghelet et al, 2003, p15; Plan, 2004, p2). Some 24 percent of the park's forest is not subject to any human intervention at all. In 16 percent of the park's forest, only minimal interventions for conservation purposes are allowed (Plan, 2004, p38). Other forest areas can be exploited, but only under strict regulations and under the park's supervision. Until 2004, the park had no strict marked boundaries. Only in 2004 was the park's territory indicated with visible markers.

Restituting forest in Dragova

The steady development of the park overlapped with the forest restitution process. In 2002, when Law 1/2000 regarding forest restitution began to be implemented, villagers received a significant area of forest within the park. The implementation of this law presupposed that a protocol would be signed between the mayoralty and the NDF. The Prefecture Pitesti notified the NDF that the transfer of forest from the state's control to the mayor's office administration had been approved. Then, the mayoralty had to restitute the land to its legitimate owners, whether private or collective, based on the request that each owner made and the documents of ownership they provided.

Some villagers simply refused to accept the forest area offered to them as restitution because they understood that they would not have complete control over their own property. It was difficult for villagers to accept that they not only had to obey the state's restrictions, but also had to accept those of the national park. One of the most common utterances I heard in the commune was: "Am I the owner of this forest or the park? If I'm the owner, I don't see any reason to obey to the park's rules. If I am not the owner then they [i.e. the state or the park] should take back this plot and give me another one somewhere else."

The next section discusses in detail how both the villagers and the park administrators have constructed and legitimized their claims.

Building and Legitimizing Claims

The conflict between the PCNP and villagers is a reflection of two competing underlying values. The park promotes "environmentalist" values, the backbone of every nature park in the world. It is enough to read the management plan to understand the aim of the park:

> *The PCNP's main aim is the protection and the conservation of some representative samples of the bio-geographic national space containing some special valuable natural elements of physical-geographic, floristic, faunologic, hydrologic, geologic, palaeontologic, soil, or of any other nature. The park offers the possibility for scientific research and educational, touristic, and recreational visits.* (Plan, 2004, p1)

The main aim of the park is not only to secure scientific or tourist activities. Its mission is also based on broader public interest. The legitimacy of the PCNP's claims is built on a national level. According to the general manager of the park (interviewed in September 2004), the state is concerned with the well-being of the wider public and therefore propagates "non-utilitarian" values. Thus, the state, as the owner of the park, would like to retain as much control as possible over its valuable natural resources.

Villagers, on the other hand, base their claims on "utilitarian" values. They argue that the forest belonged to their ancestors and that they now need it for their livelihood. The villagers of Dragova, as any other villagers in the world struggling with a national park, think on a local rather than on a national level. During my fieldwork, I never heard anyone who agreed with the park's objective of biodiversity conservation. Villagers' complaints were merely practical. "The forest is my property and I do whatever I want with it. I need the forest for my livelihood", some villagers argued. Villagers base their claims on their historical rights to the forest and on "moral economy", which is part of the livelihood argument.

Complaints come from both sides: the park complains that the villagers are stubborn and do not respect the park's land-use regulations. Two problems regularly appeared in the park's management plan and in interviews that I conducted with the park's general manager. First, private owners do not use the forest sustainably. They exploit and overgraze the park's pastures and meadows. "Regardless of the ownership status of the pastures or meadows, villagers must obey the park's rules", the park's manager stressed. Second, the villagers distrust the park's officials and their assessment of local nature, accusing the park of acting as if it is the owner of both the forest and the pasture. They believe that park officials overestimate the value of local biodiversity. The park emphasizes the national importance of local flora and fauna. Villagers, however, minimize their significance, dismissing them as "just a bloody weed" or "damn boars destroying the pastures". For local people, the forest is not as much a matter of biodiversity or landscape amenity, but rather a valuable economic asset.

Such disputes are seen, first of all, on the level of claims, since some of the villagers – namely, those who received forest in the park – question the park's history. Villagers base such claims on their historical rights to surrounding forests. While conducting fieldwork in the commune, I often heard villagers loudly protesting that "There was never a park here; our grandfathers owned these forests." Most people I spoke with thought that the park's establishment was the state's way of limiting land reform and maintaining control over the region's forests. Dragova's inhabitants dispute the property rights of the park and indirectly criticize the state's interference in their property rights.

Aside from arguing on a historical basis, private forest owners consider other state actions to be an affront to their own property rights. Small acts of defiance against the park, such as overgrazing the pastures, or illegal logging, are a response to the villager's perception that the post-socialist state is attempting to take advantage of them wherever possible. A good example of such state practices concerning the forest is the fact that a part of the forest which was restituted had been exploited by the NDF several years before the breakdown of the socialist regime. In official documents, the restituted land was described as "forest"; but it actually consists of young trees and several bushes. Needless to say, villagers were angry because they believed that the state had tricked them once again.

The two actors involved in the forest dispute use different tools to appropriate their claims. Villagers do the only thing they can do: overgraze the pastures within the park's borders and log without permission in what they consider to be their forest. The park has a powerful position because it represents the state, is managed by the Romanian Academy (and thus has public prestige), and uses high-tech tools such as satellite images to assert its claims. The satellite images establishing the park's territory with bolded borders are displayed in the mayor's entrance hall as a modern way of claiming full property rights over the land. The park's management team considers this a highly scientific method, and its prestige intimidates villagers. The park benefits from national and international support not only in terms of money, but also by its position in public discourse. For instance, a European Union program called Natura 2000 provided the park with money and facilitated international discourse on nature preservation and biodiversity conservation at the local level.

The above description would not be complete without noting that not all Dragova inhabitants are against the park. Those who own guesthouses are interested in preserving the forest because the scenery attracts visitors. Besides, a good part of guesthouse owners are outsiders and do not own forest in the area, and therefore have no reason to protest against the park. Thus, only a minority group supports the park's actions because their economic interests are better served by the park's existence. Those people who own forest outside the park are rather indifferent. The park does not bother them except when the park's wild boars destroy their pastures. They complain that they cannot shoot animals without a license. As I witnessed on several occasions, the damage caused by boars can be quite devastating. A destroyed pasture is a significant loss for stock farmers. Such farmers requested that the park pay for such damages; but the park administration has constantly refused them. "We cannot be held responsible for wild animals' behavior", the general manager told me.

Thus, villagers are not a uniform group of people struggling against the park, as each person has different interests and needs. The following section further proves this point.

Local Political Elite as Mediators

This section explores the way in which the local political elite mediate between the park and villagers, and the advantages that they obtain from this process. Among the elite who take advantage of the tense relationship between villagers and the park is the mayor. The mayor and a few local council members also take advantage of the situation by supporting the park's representatives and weakening the landowner's position. I attended a local council meeting during which the mayor asked for the council's support in changing a certain villager's property title. The park claimed that the villager had never had property rights to the plot and suggested that there was a mistake in the cadastral register. The park needed that particular plot because a specific flower called marigold (*Ligularia sibirica*) grows there and needs to be protected by a fence. The mayor asked the council members to "think green" and to protect the biodiversity in their commune because "the EU and the whole world [were] watching". "It is our duty", he claimed, "to protect our nature".[3] The mayor's intentions, however, were not as noble as his argument. It became clear that he had supported the park's proposal in order to gain the park general manager's approval for his own wood business, as shown below.

In the same line of action, the mayor's office distributed forest to several families outside the park, but in an area where some other families had collectively owned forest before nationalization. The state restituted the collective forest in 2004. Ownership documents dating back to the 1930s mention the land, but measure it in *fârtare* (a unit used in an old locally based system of measurement). The document affords 52 *fârtare* to a few wealthier families in the commune. The document also identifies the families who collectively owned the forest, but does not mention the area's boundaries. Since the restitution laws specify hectares and not old local measurements as the government's official system, decisions had to be made as to how many hectares 52 *fârtare* represent. The regional commission, located in the county capital, had the final decision regarding the land's restitution, but left the decision in the LLC's hands. The mayor, who is the head of the LLC, decided that a *fârtar* should equal 3ha and not 6ha, as the elderly members of the community claimed. In this way, he distributed forestland to new owners at the expense of other families who owned land in that area. Some of the forest owners who refused forest within the park accepted the mayor's proposal to accept forest in another location. Through this process, the mayor had satisfied both the forest owners who did not want land in the park and the park's manager, who had to deal with less stress from such "trouble-makers". From the mayor's perspective, it was a strategic move. Since this forest is outside the park, it would be easier for him to gain ownership of the forest for his logging firm in the future. Moreover, he convinced several villagers who received forest in the park to sell the land to him. He even got a better price for the land by convincing the owners that he now "has to deal with the park's regulations". However, his logging firm later obtained permission to exploit several forest plots within the park's boundaries.

The mayor plays off both the park and the villagers. As a mediator between the park and the villagers, he supports the park's claims against the "stubborn" villagers. In exchange, he supports villagers by turning a blind eye when he finds out that they cut trees illegally. The mayor promised villagers that he will get European Union funds to repair the commune's roads. He also charged himself with regaining the collective forest for the commune. In this way he manages not to be crushed between conflicting interests.

This clash between private forest owners and a national park is not unique in Romania. Other Romanian national parks have tried to enforce the same regulations on villagers, imposing a top-down approach with the same results as those described in this chapter (Vasile, 2008).

Conclusions

This chapter has explored the ways in which actors built property rights claims in post-socialist Romania. The emergence of a new national park at the same time as the restitution of forest to its former historical owners has prompted a "clash of claims" concerning property rights. Each actor bases their claims to the forest on different motivations. Villagers emphasize their historical rights and other moral aspects. The national park bases its claims on the actual property rights to the forest and emphasizes the national and international importance of nature conservation and biodiversity protection. The local political elite, who would be expected to mediate between the park and local villagers, takes advantage of this clash. The mayor and his followers abuse their political positions for personal gains.

This chapter suggests that arguments for the specification of environmental safeguards need to be complemented with a close look at the nature of the state and local power relations. The state officials in this case study, which is characterized by diametrically opposed demands by villagers and park administrators, are not the neutral arbiters envisioned by arguments for environmental safeguards. Instead, they are just as much players as referees in the "property rights game". They disregard both the villagers' demands for historical ownership and livelihood, as well as wider interests in environmental protection, in favor of their personal gains. They possess particularly powerful positions in the case study, with its stark discrepancies between the justifications underlying villagers' and the park's claims on the forest.

The argument for the inclusion of environmental safeguards to protect wider interests in the forest remains a powerful one. Yet, it needs to be accompanied by calls for an accountable state that serves local and wider interests, and not just the interests of a small political elite. As long as state officials treat forest owners as mere subjects who must obey regulations, forest owners will seek ways to circumvent environmental regulations. Yet, if parks consult local people and involve them in decision-making, villagers may be more willing to support the park's interest in preserving the forest and its biodiversity. If local state authorities are directly accountable to villagers, villagers and local government may become partners rather than rivals in the rush on forests.

Acknowledgements

Several people and institutions have contributed to this chapter in different forms. Thomas Sikor and Johannes Stahl invited me to the Towards a Rights-Based Agenda in International Forestry conference to present the paper on which this chapter is based (University of California, Berkeley, 30–31 May 2009). Jeff Romm and Nancy Peluso made my short-term post-doctoral position at the University of California (UC), Berkeley, possible. Cari Coe and Larissa Buru have critically commented on an earlier version. The New Europe College Institute for Advanced Studies Bucharest financially supported my stay at UC, Berkeley. During this time I was able to enrich this chapter. The Rachel Carson Center and the excellent LMU Munich library contributed to further improvements on this piece. Claúdia Whiteus helped to make this text readable. Last, but not least, I would like to thank the Emmy Noether Program of the Deutsche Forschungsgemeinschaft for financing my fieldwork in Dragova in 2004 and 2005. I warmly thank them all. The usual disclaimer applies here too: I am the only one responsible for this chapter.

Notes

1 This is a pseudonym.
2 During those years, anybody who owned land or forest was declared a *kulak*, and many were sentenced to either death or prison. The difference between life and death could have lain in 1ha of forest or land. Those who declared less forest than they owned hoped that they would not be included in the much-feared category of *kulaks*.
3 This episode unfolded in 2004, when the Romanian government had tough pre-accession negotiations with the EU. The environmental chapter was among the most difficult, as Romania had poor environmental politics. At the same time, the media had long-term campaigns for more sustainable and green politics.

References

Dorondel, Ş. (2005) "Land, property, and access in a village in postsocialist Romania", in Ş. Dorondel and S. Şerban (eds) *Between East and West: Studies in Anthropology and Social History*, Romanian Cultural Institute, Bucharest, pp268–307

Dorondel, Ş. (2009) "'They should be killed': Forest restitution, ethnic groups and patronage in post-socialist Romania", in D. Fay and D. James (eds) *The Rights and Wrongs of Land Restitution: Restoring What Was Ours*, Routledge-Cavendish, New York, NY, pp43–66

Hann, C. (2002) "Farewell to the socialist 'other'", in C. M. Hann (ed) *Postsocialism. Ideals, Ideologies, and Practices in Eurasia*, Routledge, New York, pp1–11

Hann, C. and the Property Relations Group (2003) *The Postsocialist Agrarian Question*, Transaction Publishers, London

Hart, G. (1989) "Agrarian change in the context of state patronage", in G. Hart, A. Turton, and B. White, with B. Fegan and L. Teck Ghee (eds) *Agrarian Transformations: Local Processes, and the State in Southeast Asia*, University of California Press, Los Angeles, CA, pp31–49

Mungiu-Pippidi, A. and G. Althabe (2002) *Secera şi buldozerul: Scornicesti şi Nucşoara. Mecanisme de aservire a ţăranului român* [*The Sickle and the Bulldozer: Scornicesti and Nucsoara. Mechanisms of Romanian Peasant Subjugation*], Polirom, Iaşi

Plan (2004) *Planul de Management al Parcului National Piatra Craiului* [*The Management Plan of the Piatra Craiului National Park*], Romania

Ribot, J. and N. Peluso (2003) "A theory of access", *Rural Sociology*, vol 62, no 2, pp153–181

Sarris, A. and D. Gavrilescu (1997) "Restructuring of farms and agricultural systems in Romania", in J. Swinnen, A. Buckwell, and E. Mathijs (eds) *Agricultural Privatisation, Land Reform and Farm Restructuring in Central and Eastern Europe*, Ashgate, Aldershot, UK, pp189–228

Sayer, A., J. McNeely, S. Maginnis, I. Boedhihartono, G. Shepherd, and B. Fisher (2008) *Local Rights and Tenure for Forests: Opportunity or Threat for Conservation?*, Rights and Resources Initiative, Washington, DC

Sikor, T. and C. Lund (2009) "Access and property: A question of power and authority", *Development and Change*, vol 40, no 1, pp1–22

Stahl, J. (2010) *Rent from the Land: The Political Ecology of Postsocialist Rural Transformation*, Anthem Press, London, New York, and New Delhi

Stahl, J., T. Sikor, and Ş. Dorondel (2009) "The institutionalization of property rights in Albanian and Romanian biodiversity conservation", *International Journal of Agricultural Resources, Governance and Ecology*, vol 8, no 1, pp57–73

Swinnen, J. F. M. (1997) "The choice of privatisation and decollectivization policies in Central and Eastern European agriculture: Observations and political economy hypothesis", in J. Swinnen (ed) *Political Economy of Agrarian Reform in Central and Eastern Europe*, Ashgate, Aldershot, pp363–398

Vasile, M. (2008) "Nature conservation, conflict and discourses on forest management: Communities and protected areas in Meridional Carpathians", *Sociologie Românească*, vol 3–4, pp87–100

Verdery, K. (1998) "Property and power in Transylvania's decollectivization", in C. M. Hann (ed) *Property Relations: Renewing the Anthropological Tradition*, Cambridge University Press, Cambridge, pp160–180

Verdery, K. (2002) "Seeing like a mayor: Or how officials obstructed Romanian land restitution", *Ethnography*, vol 3, no 1, pp5–33

Verdery, K. (2003) *The Vanishing Hectare: Property and Value in Postsocialist Transylvania*, Cornell University Press, Ithaca and London

Verghelet, M. and M. Zotta, assisted by L. Bernard (2003) *Parcul National Piatra Craiului Strategia de turism durabil* [*The Strategy of Sustainable Tourism in Piatra Craiului National Park*], www.pcrai.ro/pdf/Strategie_Turism.pdf, accessed 3 August 2010

Part V

What Political Strategies Serve Rights Recognition by the State?

Many activists share a strong emphasis on the recognition of forest people's rights by the state, as highlighted in Part IV. The chapters in Part V now look at political strategies that have proven effective for state recognition of forest people's rights, taking up the fourth critical debate highlighted in Chapter 1. Forest people and their supporters have obviously targeted national law-makers in some of their strategies, bringing about changes in national legislation with considerable success (see Chapter 2 on tenure reforms). Yet, changes in statutory law are rarely sufficient to bring about a change in forest people's actual rights. It is important, therefore, to look at the wider set of political strategies employed by forest people and their supporters to gain recognition in law *and* practice.

The following three chapters analyze various kinds of strategies that have proven effective in the recognition of forest people's rights by the state. They consider the means employed by forest rights activists, the level at which they have organized, and the specific actors and institutions that they have influenced. Their interest lies with success stories (i.e. instances of political mobilization that have eventually led to rights recognition). They do not analyze efforts that have failed, nor do they consider possible negative effects on competing claims made by other actors.

Moreover, the chapters included in this part analyze political strategies that involve some degree of organization and action above the village level. They do not look at the "everyday politics" of rights claims and rights recognition on the ground. Everyday acts of resistance to the practices of powerful outsiders and everyday contestations over rights in their material and symbolic manifestations may still remain the most important strategies employed by forest people to get their claims recognized as rights *in practice*, as indicated in various contributions to this book. Nevertheless, the focus here is on strategies that have worked for formal rights recognition by the state.

Singh shows in Chapter 13 how the level of mobilization and strategic choice of forums influence forest people's chances for recognition. She documents the actions of a group of Indian women's rights activists who successfully transpose their forest claims from local to regional levels and thereby achieve claim recognition. The activists overcome context-specific challenges by defining and redefining their claims within appropriate alternative channels. As they pursue their demands at the village level, the women are faced with negative gender-based attitudes, as well as systematic class inequalities. Changing to an alternate and appropriate political forum at the regional level, they are able to bypass negative local perceptions.

In Chapter 14, Ratner and Parnell highlight the advantages gained by activists through advocacy coalitions connecting civil society stakeholders across levels and sectors. Comparing activists' approaches in Cambodia, they find that groups employ advocacy strategies of varying efficacy to facilitate the recognition of comprehensive forest rights. Strategies that bring together relevant actors from various levels lead to more favorable legal, political, and administrative outcomes. Successful coalitions involve global and community partnerships among grassroots networks, domestic and international non-governmental organizations (NGOs), and various community groups (i.e. actors at varying levels of organization with national-, international-, and local-level support systems). Coalitions that span these multiple levels and that integrate human rights perspectives within initiatives promoting environmental management, rural development, and decentralized governance allow activists to make gains in rights recognition despite the systematic inequalities characterizing judicial and legal systems in Cambodia.

In Chapter 15, Cronkleton and Taylor highlight the significance of higher-level political mobilization for legislative recognition, but also point out the challenge of sustaining political mobilization after statutory reform. Looking at the tactics employed by collective rights claimants in Latin America, they argue that rural and indigenous groups have made significant gains with respect to legal reform movements. Examples are the rubber tapper movement in the Brazilian Amazon, an association of forest communities in Guatemala, and the federation of indigenous peoples in lowland Bolivia. Community-based mobilization and social movements affect transfers of traditional territories, allowing forest groups to secure resource management rights, rights to earn land-based livelihoods, and rights to decision-making powers. Yet, further organization and mobilization are needed post-land devolution for groups to maintain their rights to forest resources. Leaders and community members of social and collective movements are therefore challenged to maintain relevant rights once the original catalyst for advocacy has been addressed through the transfer of land rights.

Taken together, Chapters 13, 14, and 15 in this part and chapters in other parts of the book show that forest people and their supporters employ a variety of political strategies to promote forest people's claims and have them recognized as rights by the state. The strategies include national-level mobilization to lobby for legislative change (see also Chapter 10), local-level organizing to influence the implementation of national legislation and other administrative decisions (see also Chapter 6), partnerships with other civil society actors or international organizations (see also Chapter 8), and court actions (see also Chapter 3), among others. These strategies involve tactical choices about the levels of mobilization and forums used for voicing demands, extending all the way from grassroots actions to membership in global networks and appeals to transnational courts. They also involve decisions about the potential formation of coalitions across local, national, and transnational levels.

The key insights gained from the analyses, we want to surmise, is that there are many ways to promote forest people's rights, and that activists have been very pragmatic in their choice of political strategies. The large repertoire of available strategies opens up many opportunities for activists to pursue rights recognition; yet it also forces them to develop the most promising strategies for specific contexts. Thus, we end up highlighting the significance of context-specific approaches once again, just as we did in the introductory paragraphs to Parts II, III, and IV. Our conclusion does not surprise, though. If rights definitions are context specific (Part II), if the concrete actors to be supported can only be identified in relation to concrete settings (Part III), and if nested authority offers most effective recognition (Part IV), then activists have to choose their political strategies in consideration of the concrete political economic conditions under which they operate, and which they seek to modify. Universal strategies will not work.

13

Women's Action and Democratic Spaces across Scales in India

Neera M. Singh

The rights agenda in forestry includes a demand for participation of forest people in political decision-making regarding their own affairs. Forest rights activists thus demand not only redistribution of forest benefits and tenure to ensure equity, but also equitable access to decision-making space (see Chapter 1 of this volume). However, such a vision of democratization of forest governance is not easily realized as it faces a critical challenge of transforming deeply entrenched power relations at local level that lead to disproportionate appropriation of resources and decision-making space by some. Given the histories of exclusion of forest people, forest rights activists have focused more on unequal power relations between state and local communities and less on unequal power relations within communities.

In forested landscapes of developing countries, rural women depend critically on forest resources but tend to be excluded from forest governance and decision-making (Agarwal, 1997; Agarwal and Gibson, 1999; Guijt and Shah, 1998; Colfer, 2004). This chapter addresses the problem of women's exclusion from forest decision-making and illustrates strategies pursued by women in the state of Orissa in India to demand forest rights and inclusion in forest decision-making. Using a case of women's organizing within a community forestry federation, I show how women were able to use space created in the federation to advocate for their rights over non-timber forest products and expand space for their inclusion in forest decision-making at the community level. I highlight the strategies used by women and the practical implications for activists and practitioners for transforming deeply entrenched power relations. Based on the case study, I argue that power relations manifest differently across scales and marginalized actors can use emergent spaces at one scale to open space at other scales. By paying attention to these differences across scales, local actors and facilitators can leverage space at one scale to help transform power relations at another scale.

In Orissa, 8000 to 10,000 villages independently protect almost 15 to 20 percent of the state-owned forests through community-based arrangements (Conroy et al, 2001; Singh, 2002). Orissa is one of the most forest-rich and economically underdeveloped states in mainland India. Almost 40 percent of Orissa's geographical area is categorized as forests, while actual forest cover is about 23 percent. A majority of the rural population lives in forested landscapes and depends on forests and marginal lands for subsistence and livelihood needs. This forest dependence has led villagers to invest in institutional arrangements to collectively protect and manage state-owned forests even in the absence of legal rights. While some of the community-based forest management initiatives date back to the 1930s, most others started during the mid 1980s in response to forest degradation (Kant et al, 1991). Surprisingly, the state has not built up on these initiatives. It has mostly ignored them in the past, and in recent years has sought to co-opt them under the Joint Forest Management (JFM) program. In response to the continuing state apathy for local communities' demand for legal rights over the forests that they have protected, the forest-protecting villages have formed federations at different levels (Singh, 2002). A state-level federation called Orissa Jungle Manch (OJM) was formed in 1999. As of 2009, there were 25 district and other local-level federations in Orissa.

While, at one level, these community forestry initiatives represent efforts to democratize forest governance and challenge state autonomy in forest decision-making, they are not very democratic internally and marginalize certain groups of people, particularly women, *dalits*, and *adivasis* (Sarin et al, 2003).[1] Community forestry federations also tend to be male dominated as women's exclusion at lower levels is amplified in federations where representation at higher levels is contingent on presence or absence of minorities at lower levels. In the case presented here, there has been an engendering of the community forestry federation, and this chapter analyzes the process through which space has been created for women and how women have used that space for opening opportunities at other scales.

The Case Setting: Maa Maninag Jungle Surakhya Parishad (MMJSP)

MMJSP is a federation of forest-protecting villages in Ranpur block[2] of Nayagarh District in Orissa. In the subsistence agriculture-based economy of

Ranpur, forest-based livelihoods play a critical role in the lives of the poor, especially of *adivasis* and *dalits*. Almost 35 percent of the geographical area of the block is recorded as forest. About 39 percent of the households are landless, and in the case of *adivasis* and *dalits*, it is as high as 70 percent. In general, rural dependence on forests is high and more so for the landless. This dependence led villages to respond to forest degradation by taking the initiative to protect forests. Of the 271 village settlements, 187 have been conserving forests for the past 20 to 30 years.

MMJSP is named after a local deity and its namesake hill, Maa Maninag, and translates as "Mother Maninag Forest Protection Forum". It is locally referred to as the Parishad. It was formed in 1997 with an initial membership of 85 villages. In 2009, its membership stood at 187 forest-protecting communities clustered into 18 groups, based on historic social, cultural, and shared resource-related ties, with membership ranging from 4 to 14 villages.

MMJSP and Space for Women

MMJSP, like other federations, was initially dominated by men. No village women were present at the initial local meetings that led to the formation of MMJSP. Vasundhara, a Bhubaneswar-based non-governmental organization (NGO), facilitated the formation of MMJSP and has been instrumental in nudging MMJSP to create space for women. At the initial meetings of MMJSP, the male leaders vehemently opposed Vasundhara's suggestion of providing space for women's representation within MMJSP's governance structure. They finally agreed to include women in the governance structure; but this was done to appear gender sensitive and progressive to outsiders (Singh, 2007).

During 1997 to 1998, women's presence in MMJSP remained marginal (Singh, 2009). To address this problem of lack of women's involvement, a women's sub-committee or a task force was formed within MMJSP to design strategies for improving women's involvement in the federation. In 1999, soon after its formation, this task force organized a series of village-level meetings with women to discuss the constraints to women's participation in MMJSP. At these meetings, women suggested that women meet separately at the block level on a regular basis. On 26 September 1999, the first "women's meeting" was convened. At this meeting, participants decided to form a women's sub-group called the Central Women's Committee (CWC) and agreed to meet every month on the 18th day of the month. Women call these meetings "*mahila* meetings", or women's meetings, or simply "*athraha tarikh* meetings".[3] Even though women's presence was low at the initial meetings, the numbers soon increased due to the creation of the women-only space within MMJSP.

Space for women through the "*athraha tarikh*" meetings

The "*athraha tarikh*" meetings served as an "open space" for women to discuss their problems, meet other women, and learn from each other's experience.

While in the initial meetings, the male leaders of MMJSP tended to dominate, since 2000, a tribal woman has presided over these meetings and women have started treating the space as theirs. The male leaders still attend and their presence helps to maintain linkages with MMJSP, but also sometimes hinders discussion.

The meetings are held in Ranpur town in the shared office premises of MMJSP and Vasundhara that is centrally located and easily accessible, and last for about three hours. Initially, the middle-aged and older women were a majority at these meetings, given their relative freedom from household responsibilities and lesser constraints on travel. During recent years, younger women have started coming as the CWC has begun addressing forest-based livelihood issues that concern them.

As women gathered every month, they brought their livelihood problems to these meetings and to the notice of MMJSP. In the process, women's livelihood issues have been added to MMJSP's advocacy agenda. The following section discusses women's action and advocacy relating to rights to secure livelihood from the gathering of kendu leaves, a non-timber forest product whose trade is nationalized in most parts of India.

Rights over Non-Timber Forest Products: Women's Advocacy Struggle

Leaves of the kendu tree (*Diospyros melanoxylon*) are used for wrapping *beedis* (the local cigarettes). Collection of kendu leaves is an important source of livelihood for the poor in central India. It provides much needed cash to the poor, especially during the lean summer months when few other sources of employment exist. In Orissa, about 30 million person days of work are created in the collection of kendu leaves within a short span of three to four months. Kendu leaf trade in Orissa is nationalized; hence, kendu leaves can be sold at only government-run collection centers called *phadies*. In Ranpur, there were no *phadies*, and women gathering kendu leaves were forced to sell to private traders who operate illicitly and offer only a fraction of the government-fixed prices. Women raised this problem at one of the women's meetings in 2000. After discussing the issue at at the *athraha tarikh* meeting in January 2001, women decided to organize a rally in Ranpur town to demand the opening-up of kendu leaf *phadies*. In April 2001, approximately 2000 women from 95 villages rallied around the demand for *phadies* and sent a petition to the chief minister of Orissa. In response, in 2002, the government established two *phadies* and promised to open more *phadies* later. However, the government did not deliver on this promise.

Women continued advocating for additional *phadies* through regular petitions. When these efforts failed, women decided to organize another rally in Ranpur. In November 2004, more than 1500 *dalit* and *adivasi* women gathered for this rally. After this rally, women decided to do a sit-in demonstration (*dharna*) in front of the State Legislative Assembly. From 9 to 16 March 2005, 19 women sat in a *dharna* in front of the assembly in Bhubaneswar.

The process of advocating for kendu leaf *phadies* has taken women into territories and spaces that they usually do not inhabit. Travelling to these places and spaces, both physically and metaphorically, has been liberating and empowering for women. This has also made them see and realize their power and potential as political actors.

For women, their action for kendu leaf *phadies* was an important turning point. In my interviews, they recounted this struggle, and saw their first rally at Ranpur in 2001 as an important marker in their becoming political actors. At a meeting in 2004 to discuss future courses of action, women were fast to suggest: "*Puni rally kariba*" ("let us do another rally"). The women who had benefited from the first kendu leaf *phadi* felt compelled to continue with advocacy efforts. They said: "Because of everyone's efforts, we got a *phadi*. How can we now sit silent?"

Two women *sarpanches*[4] were also actively involved in women's meetings. They brought to these meetings their experience from the formal political arena. This advice, as well as counsel from male leaders of MMJSP, helped women to design multi-pronged strategies. For many of the women who went to Bhubaneswar for the *dharna*, it was their first visit to the capital city. Being in alien spaces and places and finding their feet was in itself an experience. There was also the rare freedom from the drudgery of household work. The "streets of power" in Bhubaneswar thus became their training grounds in the arena of political action and activism (Singh, 2009).

The representative to the state legislative assembly from their region was a woman who also held a ministerial charge. Women were particularly hurt by the apathy of this female political leader during the *dharna* process. At the time of seeking votes, she had invoked "sisterhood" when asking for women's votes and had claimed to understand women's problems. However, this *Apa* (elder sister) "changed" after getting elected; and the women felt cheated at the lack of response from her as they sat on *dharna* in Bhubaneswar. After six days of *dharna*, the women went to meet this female member of the legislative assembly (MLA). In the interaction that ensued, Kuntala Nahak, a *dalit* woman, took on the female minister, and reprimanded her in very strong words for not doing anything about their problem (Singh, 2007). Kuntala, in many ways, was not awed by her minister status or limited by the demands of correct etiquette, and was able to speak to power directly and demand accountability. After this encounter, the MLA met with the women representatives from MMJSP again and promised to open another *phadi* in Ranpur. While this fell short of women's demand for several *phadies*, the women felt that they were not returning home empty handed.

Other than the advocacy for kendu leaves, women have used the emergent space within MMJSP to organize around other forest-based livelihood issues. Biskia Jani, an *adivasi* woman leader, says: "When we started meeting, we thought only forest protection is not enough … unless we can get some income from the forests. So, we looked at our existing activities … for example, siali leaves – and decided to improve our incomes from siali." Siali leaf (*Bauhinia vahlii*) is an important non-timber forest product (NTFP) of central and southern Orissa used widely by grocery stores, local restaurants, and for packing material. Women decided to obtain machines for stitching siali leaf plates and market them collectively. This collective action was effective at a larger scale, and women were able to act at this scale as a result of organizing through MMJSP.

Women's action within MMJSP is leading to transformation in the democratic space available to them at other scales. The following section discusses the transformations in women's participation and space available for participation at different scales.

Transformation in Spaces across Scales

At the regional scale within MMJSP

Through their advocacy efforts and action in emergent spaces within MMJSP, women have displayed their power and determination to take control over their lives. In the process, the men have come to appreciate the women's role and, more importantly, their power as mobilized masses that can rally in the streets and display MMJSP's strength.

Over the years, there has been an increase in women's representation in the governance structure of MMJSP, which consists of a general body, an executive committee, and a working body. While the executive committee is the main decision-making body, a larger "working body" was created to involve more people in the various roles of MMJSP, such as conflict resolution, policy advocacy, and lobbying. In addition, there is an advisory committee, a special task force for conflict resolution, and the Central Women's Committee. Initially, the general body consisted of the president and secretaries from all the village-level committees. Given the male domination at the village level, no women were part of the general body. In 2000 the institutional norms were changed, and the entire executive committee of cluster-level groups was made to be a part of MMJSP's general body. This allowed women to come in. This institutional change shows how space can be created by favoring one scale over the other. If community scale had continued to define membership criteria to higher-order organizational structures, then women would not have any scope and space to participate.

The executive committee of MMJSP is usually elected by the general body. With an all-male general body, normally the executive committee would not include any women. These changes in MMJSP's governance structure were due Vasundhara's interventions to increase women's involvement in community forestry in the area. The initial *ad hoc* executive committee of MMJSP in 1997 consisted of 24 men. At the time of MMJSP's registration as a society in

1998, 3 women were included in the 11-member executive committee. These three women included one *dalit* and two *adivasi* women. This inclusion initially remained notional, but has been changing over the years. The current executive committee of 17 people has 4 women; however, until early 2006, none of the women were office bearers.

A working body was constituted to increase direct participation from a larger body of villagers in the functioning of the MMJSP. The initial working body in 1997 consisted of 30 people, with only 1 woman member. The size of the working body and the proportion of women have grown steadily. By 2006, the working body had 84 members with 31 women. Thus, there has been a constant increase in the membership of women in the working body. The representation of *adivasis* and *dalits* has also been increasing. This is in stark contrast to membership of women at the village level. According to a survey conducted by Vasundhara and MMJSP in 2005, in the 111 villages involved in forest protection, only 23 villages had women representatives on the forest protection committees. Only 7 percent of the office bearers in these committees are women. Representation of *dalits* and *adivasis* is equally low at 7 percent and 16 percent, respectively. Women's actual presence and participation in the committee meetings tends to be minimal.

Even though the CWC started as a separate women's group, there is synergy between men's and women's actions. As one of the women leaders said: "If we [women] are in the fore, they are behind [us]. And if they are in the fore, we support them. When we inform the men about a problem, they cooperate with us to solve it." The women also feel that they now have more respect. As Kuntala said: "Earlier, when we used to speak they did not listen. Now, things are different. Our opinion is heard [*katha suno chanti*]."

At the community level

In Ranpur, as is typical in coastal Orissa, patriarchy is strong and women tend to be excluded from community decision-making, including community forestry. Various studies indicate that women faced hardship during the initial years of forest protection when their access to forests was restricted (Sarin, 1998; Agarwal, 2001; Singh, 2001). Although the situation improves as forests regenerate and women's access to forests is often restored, their exclusion from forest decision-making continues.

During recent years, many villages have "put" women on the village forest protection committees due to the institutional provision for women's representation on the JFM committees. However, their inclusion remains notional and they hardly attend the committee meetings, or if they sometimes do attend, they sit in and do not participate (cf. Nightingale, 2002). Men decide when to meet and according to the women "never inform or invite us ... We would go if we were invited. How can we go uninvited?" As is widely noted in the literature, meetings are scheduled without taking into account women's time availability or convenience.

Women's exclusion at the community level is due to strong social and cultural taboos that restrict their participation in the public sphere. These cultural constraints also get in the way of women acquiring the skills needed for public participation, and reinforce their marginalization. Women frequently also point out constraints related to confidence. In my interviews, women said: "If we go once, we will get the courage; [but] we have never gone. How can someone who has never opened her mouth speak? Once she speaks up, then she can."

At the community level, women have great difficulty in breaking the deeply entrenched cultural taboos. As one woman told me: "Someone [at the meeting] is a father-in-law or uncle-in-law; how can we sit with them ... won't we feel shy? Men will say: how immodest she has become. We have to take care of our honor." Under these circumstances, women also have very little opportunity to acquire the skills and confidence for public engagement.

At the village level, a woman is entrenched in her identity as someone's daughter-in-law, mother, or sister-in-law, while these identities and subject positions become more fluid at another scale. Cultural taboos are also more relaxed outside of the immediate village boundaries. By organizing at a higher spatial scale, women were able to acquire skills that they cannot gain at the community level, and were able to circumvent cultural constraints that hinder their participation at the village level. Some of these women were later able to use these skills to demand their right to be included in forest decision-making at the community level.

For example, the *dalit* woman, Kuntala, who confronted the female minister in Bhubaneswar, refused to sign a village resolution without reading it after this experience, and demanded changes in it. Her success in engaging with democratic politics at a higher scale gave her both the prestige and the confidence to expand her role at the community scale.

During recent years, after women's involvement in MMJSP, there have been several instances where women have taken defunct forest protection systems over from men. In Dengajhari village, women revived community forest protection systems after several years of neglect. After attending a CWC meeting on 18 October 2000, women decided to revive the forest protection system, with two groups of women patrolling on a rotational basis. Soon after they started protecting the forests, the women faced a serious confrontation with a timber smuggling group. When a group of 50 to 60 men tried to steal timber from their forest, 27 women confronted them and seized more than 1500 logs. The village men, who were too afraid to confront the timber smugglers, were amazed at the women's courage, and this incident became a local legend. The leader of the women's group came to be known as forester *mausi* (aunt) and her story inspired women in the neighboring villages to play a greater role in forest protection. Men also became open to the possibility of a more pronounced role for women in forest protection. There are now several instances where women have started protecting forests on their own.

Women members of MMJSP committees are also beginning to question the exclusion of women at the community level, more generally. In the general body meeting of MMJSP in 2007, Pramila Dash, of Surkhabadi village of the Das Mauja cluster, raised the contradiction between the rhetoric about women's participation at the block level, and the actual situation at the village-level reality where women are simply expected to provide their signatures to endorse men's decisions. Thus changes are occurring both because women who have achieved success at a higher level are beginning to stand up at the community level, and because actors at the block level, where greater democratic space already exists for women, are lobbying for institutionalizing the same changes at the village level.

Continuing challenges at other scales

While there has been a transformation in women's participation in MMJSP, women's involvement in other community forestry federations remains a challenge. This is not unusual considering that the community forest management groups are traditional institutions that derive strength from culture and traditional sources of authority. Their federations also draw on these strengths of tradition and culture, and often they do not know how to address power inequalities that come with this tradition and culture. In Orissa, only 3 of the 25 federations have taken steps to improve women's involvement. Most federations do not necessarily see marginalization of women or other social groups as a problem. While NGOs working with federations may recognize this as a problem, they do not necessarily allocate their scarce resources to processes of long-term change.

The institutional norms of representation in federations tend to promote leaders or office bearers from lower to higher scales. As women are absent from governance structures at the community level, this absence persists at higher federation scales, unless some corrective measures to specifically include women are taken at other scales. Even when women are included through special provisions, they do not have the same representational authority as the male "leaders".

In Orissa Jungle Manch (the state-level federation), women's involvement and participation remains marginal. In addition to having a limited institutional presence, women also remain absent from different state-level policy forums, conferences, and workshops. Many factors contribute to this. Often it is leaders and those seen as "experts" who attend policy forums. Women are not able to attain such leadership and expert positions due to limited opportunity to do so. When district federations nominate one or two representatives to represent the district at the state level, they tend to be men. For village women, travel to Bhubaneswar is also a constraint as their families often insist on a female travelling partner. This constraint, and the travel costs of each additional person's travel, favors men's representation in state-level policy events.

When women do come to these meetings, they find it hard to comprehend policy discourse and discussion of issues that seem far removed from their immediate concerns. Men often do not take the time to fill in the details of past discussion. Women also encounter difficulties comprehending technical and managerial discourse.

Conclusions

This chapter has focused on strategies for addressing the problem of women's marginalization in forest resource governance and in facilitating women's demand for their right to be included in forest decision-making. In societies where hierarchy and unequal social relations define the prevailing social order, top-down approaches to social reform do not work very well (Cornwall et al, 2006; Batliwala and Dhanraj, 2006; Mukhopadhayay, 2006), and bottom-up processes for social transformation are unlikely to emerge from locales where power relations are strongly entrenched. This case study suggests that it might be productive for practitioners and activists to focus on locales and spaces from which change can come.

Constraints and possibilities for women's participation are different across spatial scales, and it is important to better comprehend and leverage these differences. While cultural constraints restrict women's voice within their immediate communities, they are often more knowledgeable about the workings of this scale. At the macro-scale of region, nation, or globe, techno-managerial discourse and language make it more difficult for village women (or men) to participate effectively. In Orissa, meso-scales (i.e. scales of locality and regions where common struggles of life and livelihood can be clearly identified) provided an easier entry point to addressing issues of marginalization. In the case of MMJSP, women were able to gain a voice at the block level, as that was the scale at which they needed to take up advocacy to solve their commonly faced forest rights and livelihood problems. This meso-scale also provided a nurturing ground to build confidence and to initiate women into the policy arena and public domain.

Even though the concept of scale and cross-scale linkages has drawn a lot of attention in the social sciences during recent years, there has been inadequate appreciation of how democratic spaces across scales differ, and the practical implications of this for social organizing and rights-based social movements.

I suggest that closer attention to constraints and possibilities at different scales can help practitioners and development workers to design strategies that are able to leverage spaces at meso-scales, while supporting spillover effects at other scales. This requires recognition of interrelationships across scales, as well as enhanced communication and connections.

This case study shows that it is important to support the "jumping" or traversing of scales by marginalized people as a way of facilitating social transformation and correcting power imbalances. This allows ideas to flow across scales, connections to be made, and skills to be gained at one scale and applied at another where it would be difficult to acquire them.

Notes

1 *Dalit* refers to a caste traditionally regarded as untouchables. *Adivasis* are believed to be the aboriginal population of India and are also referred to as "tribals". Both groups tend to be discriminated against in rural India.
2 A block refers to a district sub-division used for administrative planning purpose. Under the Indian self-governance system (i.e. the Panchayati Raj), there are three tiers of governance: *gram panchayat* (cluster of villages), block, and district.
3 *Athraha* means "18th" and *tarikh* means "date" in Oriya, and refer to the meetings held on the 18th of every month.
4 A *sarpanch* is a democratically elected head of a village-level statutory institution of local self-government called the *gram panchayat* in India. In 1993, through the 73rd amendment of the constitution, one third of all *panchayat* seats have been reserved for women.

References

Agarwal, B. (1997) "Environmental action, gender equity and women's participation", *Development and Change*, vol 28, no 1, pp1–44
Agarwal, B. (2001) "Participatory exclusions, community forestry, and gender: An analysis for South Asia and a conceptual framework", *World Development*, vol 29, no 10, pp1623–1648
Agrawal, A. and C. C. Gibson (1999) "Enchantment and disenchantment: The role of community in natural resource conservation", *World Development*, vol 27, no 4, pp629–649
Batliwala, S. and D. Dhanraj (2006) "Gender myths that instrumentalize women: A view from the Indian front line', in A. Cornwall, E. Harrison, and A. Whitehead (eds) *Feminisms in Development: Contradictions, Contestations, and Challenges*, Zed Books, London and New York, pp21–34
Colfer, C. J. P. (ed) (2004) *The Equitable Forest: Diversity, Community, and Resource Management*, Resources for the Future, Washington, DC
Conroy, C., A. Mishra, A. Rai, N. M. Singh, and M. K. Chan (2001) "Conflicts affecting participatory forest management: Their nature and implications", in B. Vira and R. Jeffrey (eds) *Analytical Issues in Participatory Natural Resources Management*, Palgrave Macmillan, New York, pp165–184
Cornwall, A., E. Harrison, and A. Whitehead (2006) *Feminisms in Development: Contradictions, Contestations, and Challenges*, Zed Books, London and New York
Guijt, I. and M. K. Shah (1998) *The Myth of Community: Gender Issues in Participatory Development*, Intermediate Technology Publications, London
Kant, S., N. M. Singh, and K. K. Singh (1991) *Community Based Forest Management Systems: Case Studies from Orissa*, Swedish International Development Authority, ISO/Swedforest, and the Indian Institute of Forest Management, New Delhi and Bhopal
Mukhopadhyay, M. (2006) "Mainstreaming gender or 'streaming' gender away: Feminists marooned in the development business", in A. Cornwall, E. Harrison, and A. Whitehead (eds) *Feminisms in Development: Contradictions, Contestations, and Challenges*, Zed Books, London and New York, pp135–149

Nightingale, A. (2002) 'Participating or just sitting in? The dynamics of gender and caste in community forestry', *Journal of Forestry and Livelihoods*, vol 2, no 1

Sarin, M. (1998) *Who Is Gaining? Who Is Losing?: Gender and Equity Concerns in Joint Forest Management*, Society for Promotion of Wastelands Development, New Delhi

Sarin, M., N. Singh, N. Sundar, and R. Bhogal (2003) "Devolution as a threat to democratic decision-making in forestry? Findings from three states in India", in D. E. and E. Wollenberg (eds) *Local Forest Management: The Impacts of Devolution Policies*, Earthscan, London and Sterling, VA, pp55–126

Singh, N. (2001) "Women and community forests in Orissa: Rights and management", *Indian Journal of Gender Studies*, vol 8, no 2, pp257–270

Singh, N. (2002) "Federations of community forest management groups in Orissa: Crafting new institutions to assert local rights", *Forest, Trees and People Newsletter*, vol 46, pp35–45

Singh, N. (2007) "Transgressing political spaces and claiming citizenship: The case of women kendu leaf-pluckers and the Community Forestry Federation, Ranpur, Orissa", in S. Krishna (ed) *Women's Livelihood Rights: Recasting Citizenship for Development*, Sage Publications, New Delhi, Thousand Oaks, and London, pp62–81

Singh, N. (2009) *Environmental Subjectivity, Democratic Assertions and Re-imagination of Forest Governance in Orissa, India*, PhD thesis, Michigan State University, East Lansing, MI

14

Building Coalitions across Sectors and Scales in Cambodia

Blake D. Ratner and Terry Parnell

Because of their importance to local livelihoods, natural resource rights have become a focal point for the most active experimentation and evolution of human rights-based advocacy in Cambodia. Human rights, community development, and conservation organizations engaged in supporting communities to defend resource rights face severe obstacles. These include a rapid and broad reallocation of state-owned and privately held lands to commercial concessions, widespread local land conflicts, a court system that has proven systematically ineffective at resolving such conflicts independently, and a fast-changing legal and regulatory framework. Civil society organizations have adopted a wide range of approaches to seek progress amidst these obstacles, learning from each other, building alliances, and adapting through experience. And yet, while there are notable advances, many engaged in this work feel that the collective efforts of civil society are too fractured, their results falling short of potential, and that a more coherent approach is needed.

Based on a comparative analysis of efforts by community groups, as well as domestic and international non-governmental organizations (NGOs) working to assist forest-dependent communities, this chapter explores the practical efficacy of a range of rights-based approaches in securing equitable forest rights and reducing conflict. The demands of rights-based advocacy in Cambodia combine the three elements explored in this book. Local forest-dependent communities are seeking equity in the distribution of forest benefits, primarily through recognition and defense of tenure rights; they are seeking affirmation of forest people's identities, experiences, and distinct visions of development, most notably in the case of the indigenous peoples who live predominately in the country's historically forested mountain regions; and they are seeking participation in political decision-making, especially to counteract what most local residents view as a rapid and unwelcome intrusion of commercial interests in forest areas. Our focus in this chapter is the strategies that civil society groups have adopted to advocate for and defend forest rights, probing the role of different types of civil society groups, the links among them, the methods they employ, and the difference that coalitions make in advancing the forest rights agenda in the face of substantial opposition. Specifically, the chapter compares approaches, emphasizing:

- assertion of rights by engaging with authorities, seeking legal or administrative recourse, direct action, and mediation;
- claiming tenure for decentralized community-based natural resources management through development of community forestry; and
- advocacy of broader economic and social rights, including the rights of ethnic minorities to self-determination and local participation in decision-making.

Because implementation of the legal and judicial framework is so flawed, the ability to align the actions of multiple stakeholders, domestically and internationally, is a key factor influencing the efficacy of these approaches. In particular, this entails a more coherent integration of human rights perspectives within initiatives promoting environmental management, rural development, and decentralized governance, and it entails investing in capacity for collective action to support ecosystem-based management beyond the local scale.

The chapter is organized as follows. The next section outlines the recent history of forest rights and conflict in Cambodia to provide a context for understanding the constraints facing civil society organizations and the approaches that they have adopted during recent years, particularly after the suspension of logging concessions and the emergence of a legal framework for community-based forest management. Three approaches are then described and illustrated, drawing on interviews with leaders from grassroots networks, and domestic and international NGOs. The subsequent section analyzes the advantages and limitations of these approaches, and provides recommendations for enhancing the value of human rights levers in improving forest governance and livelihood outcomes for forest-dependent communities. The conclusion discusses the implications of Cambodia's experience in international perspective.

Forest Policy, Resource Conflict, and Local Livelihoods in Cambodia

Forest logging financed armed conflict during Cambodia's protracted civil war, particularly during the decade from the withdrawal of Vietnamese forces in

1989 through the dissolution of the Khmer Rouge in 1998. Even as the armed conflict subsided, however, logging operations remained a critical source of revenue for competing political groups, including the former opposition factions who had been integrated within the coalition government through United Nations-brokered elections in 1993 (Le Billon, 2000; Le Billon and Springer, 2007). By the mid 1990s, timber exports accounted for an estimated 43 percent of export earnings, more than any other country in the world, according to the Food and Agriculture Organization of the United Nations (FAO) (Le Billon and Springer, 2007). Despite a series of temporary bans on logging and timber exports, commercial exploitation of the country's forest resources continued. This not only brought serious losses to the environment and local livelihoods, but also undermined the significant international investment in promoting institutions of democratic governance in the country (World Bank et al, 1996; Talbott, 1998).

The response of international aid agencies, led by the World Bank, focused on improving regulation of the commercial forest sector. Measures included capacity-building to strengthen the technical oversight provided by forestry authorities, to increase compliance of concessionaires with the terms of their contracts, and to collect official revenue from these operations, combined with independent forest monitoring through a Forest Crime Monitoring unit staffed by Global Witness, an international human rights NGO. These measures did constrain illegal logging, and ultimately all commercial forest concessions have either been cancelled or suspended for failure to comply with regulations, including the requirement to produce a forest management plan. Notably, community and NGO advocacy was instrumental in compelling the termination of the concessions (Shadravan, 2003).

Yet, new legal mechanisms for allocating private tenure or resource extraction rights are evolving rapidly. These include economic land concessions intended for agro-industrial development, typically plantations for export crops, and "exploitation, development and use" concessions offered for a variety of purposes, of which mineral exploration or extraction is most common. These newer forms of concessions are less demanding for concessionaires in terms of restrictions on resource use; their allocation is also less regulated, and while at least some requirements for resource management are stipulated in law, enforcement remains weak. In many instances, economic land concessions or exploitation, development, and use concessions are valued because they offer an indirect route for companies to clear forested areas, or to establish tenure rights to the land itself, often in opposition to the claims of local communities. Concurrently, new laws have also opened legal space for community-based natural resource management, part of a broader policy promoting decentralized governance. While these legal reforms offer a route to establishing community-based management, they have not erased the underlying dynamics of resource competition, nor have they yet fundamentally shifted, in practice, the distribution of economic benefits from the forest sector at significant scale.

Today, most forest-dependent communities throughout the country find themselves in tenuous territory: by law they are guaranteed a range of protections against resource expropriation and a route to formalizing community rights to use and manage forests; in practice, they are highly vulnerable and many are struggling to retain access to resources that are a pillar of their livelihoods and critical to their food security. Often, too, they face adversaries whom they do not understand: companies that have been granted tenure or resource extraction rights through a process that is completely opaque to the communities affected.

Ideally, good governance – including rule of law and effective mechanisms of accountability for both public and private power holders – should be established first if commercial resource extraction is to contribute to economic development, equitable distribution of benefits, and poverty reduction (Davis, 2005). In Cambodia today, groups sympathetic to these goals face the much more challenging task of promoting improvements in governance in the midst of widespread resource conflict. There are no simple answers to the question of how to achieve progress towards good governance so that forest and land conflicts can be resolved equitably, and local rights protected, while preserving and rehabilitating the quality of natural resources to provide a long-term basis for people's livelihoods. The next sections detail the experience of international and domestic NGOs and civil society networks with three broad approaches.

Three Approaches to Promoting Community Resource Rights

The three approaches are distinguished not by the actors involved but rather by the focus of their attention. They may focus on:

1. assertion of rights by engaging with authorities, seeking legal or administrative recourse, direct confrontation, and mediation;
2. claiming tenure for decentralized community-based natural resources management through development of community forestry; or
3. advocacy of broader economic and social rights, including the rights of ethnic minorities to self-determination and local participation in decision-making.

These are not mutually exclusive approaches; indeed, some organizations are directly pursuing or indirectly supporting efforts at all three.

Community empowerment through mobilization, direct action, and mediation of land conflicts

Efforts at community mobilization in defense of local people's land rights or access to forest resources are often spurred by immediate threats, resulting from a concession or other major land claim, or illegal logging or poaching. Direct non-violent action has been successful in slowing land clearing, gaining the release of detained activists, and securing negotiations in a number of cases. It has also brought land and forest conflicts to the attention of the broader community. As one grassroots leader explained:

> *As poor people, we do not have much power by ourselves and it is easy to ignore us. When we stand together, the government, the company, the media, and other Cambodians are more likely to pay attention to us. Blocking a bulldozer with our bodies may be dangerous, but we do this because we know that the loss of our land and livelihoods is even more terrible. People say: "We would rather die by a bullet, than by starvation." So when we protest it is the ultimate way of saying: "Do not ignore us. We are citizens and we have rights too."*

Except in cases where threats arise suddenly, direct action is rarely undertaken without communities attempting a variety of other avenues first. This includes seeking assistance from local authorities, attempting to talk with company managers and workers, petitioning various departments, ministries, and elected officials from the local level up, and filing complaints with the courts. One of the reasons communities revert to protests is that pursuing individual cases through formal court or complaint processes is widely recognized as ineffective. As explained by Kit Touch, an advisor with the Community Legal Education Center (CLEC):

> *The law says that people can sue, can protest, can express their opinions openly, and the courts will protect villagers in exercising these rights; but in reality the court does nothing but serve personal interests. If a wrongdoer bribes the judge, the judge will protect him, if the rich or powerful put pressure on the court, the judge will comply. Therefore, villagers cannot rely on the courts.*

Understanding the limitations of the courts, however, does not mean dispensing with the law as a route to legitimizing community tenure and access rights. CLEC's Indigenous People's Access to Justice Project, for example, provides legal services to communities to gather evidence and build a legal case in their defense, then seeks resolution outside of the courts. ADHOC, one of the longest-standing human rights NGOs in Cambodia, has traditionally focused on bringing individual cases to media attention and to the courts, along with legal education for communities facing conflicts. Pen Bonnar, head of the ADHOC office for Ratanakiri, describes how the group achieved very little success after years of individual court cases, as well as subsequent workshops to raise awareness among authorities about disputes over community forestland. He decided that to make progress, ADHOC would have to organize demonstrations and bring the national spotlight on land conflicts affecting minority communities in Ratanakiri. While many disputes in the province remain unresolved, the approach has succeeded in gaining recognition from the highest levels in government. "The people involved are concerned", says Pen, "because they know these are very serious violations of the law."

Working with government to enable community forestry

A second approach focuses on supporting community-based forest management, which includes efforts to strengthen the legal framework for community forestry, and to secure resource management rights and benefits for communities as provided under that framework. While many forest-dependent communities in Cambodia have maintained traditional arrangements for allocating access and use rights for a range of forest resources within and among villages, the legal foundation for community forestry emerged recently. While the 2002 Forestry Law references decentralized natural resource management and customary rights of communities to use forest resources, the implementing regulations to give communities a roadmap towards official legal recognition of community forest areas were only adopted in 2006.

Even with this roadmap in place, however, the route for communities to obtain full legal recognition and, ultimately, to secure a long-term flow of benefits from community-managed forest areas is fraught with obstacles. As Sunderlin (2005) notes, the official policy commitment to community forestry has been ambiguous, and the prospect of community forestry contributing to secure livelihoods and poverty reduction is undermined, in practice, by government policies favoring the military, private concessions, and rent extraction.

Recognizing these constraints, a number of organizations focused on community forestry have developed activities to support communities in fending off competing resource claims in advance of, or in tandem with, their efforts to establish officially recognized management areas. The Regional Community Forestry Training Center (RECOFTC), for example, has developed a mediation approach, employing trained local mediators to intervene, sometimes at the point when community members have already gathered to confront bulldozers hired by the concessionaire, then working with the Forestry Administration to have them open negotiation with the concessionaire on behalf of the communities. Explaining the rationale for such targeted interventions in the case of a conflict at Kbal Damrei in Kratie Province, Bampton et al (2008) highlight unequal access to decision-makers. It is clear that the procedures for granting the economic land concession have not been followed properly, although at the same time the villagers are striving to complete all the requirements for establishing and obtaining legal recognition for their community forest according to the law, but clearly without sufficient support from the authorities who should be supporting them.

Where legal recognition of community forestry is most advanced, there are some signs that this may contribute to the ability of communities to fend off competing land claims. As of July 2010, 145 sites have obtained 15-year community forestry agreements approved by the Ministry of Agriculture. To date, none of these agreements has ever been challenged or defended in court (Bradley, 7 May 2009, interview); yet, in some places the legal recognition granted by community forestry agreements appears to be offering a degree of tenure protection in practice:

> *According to Mr. Prak Marina, Siem Reap Forestry Administration Cantonment Deputy ... "When buyers come, the first question they ask is: 'Is the land under community forestry?' Buyers want to avoid the burden of community consultation and the risk of conflict, which could derail their purchase. As a result, according to another Forestry Administration officer, Mr. Kung Boravuth, community forestry areas are becoming like 'forest islands' since the surrounding land is quickly being sold off and converted."* (Bradley, 2009, pers comm)

Yet, it is important to note, too, that sites approved through community forestry agreements are, by definition, less conflictual: one of the steps towards recognition is to establish that there are no competing claims to the land. Therefore, the areas where community rights are most disputed may be left out of the approval process. Another concern is that as economic concessions expand to accommodate a growing agro-industry, community forests and other sanctioned community areas are challenged by concessionaires with competing land claims. A wide range of such cases have already been documented (IPNN, 2010).

Phan Kamnap, chief of the Community Forestry Office of the Forestry Administration, readily acknowledges the multiple threats facing forest-dependent communities and argues that establishing legal rights for community management is one of the best tools available to protect their livelihoods. Noting the limited influence of the Forestry Administration over decisions such as the location of economic land concessions, Phan argues that communities need all the help that they can get. "NGOs are important partners of government", he says. "If not for them, the government would have more difficulty implementing the policy ... The legal framework is in place to support community forestry. What we need to make it work is to deliver the social and economic benefits."

Broader human rights advocacy

A third approach focuses on networking and advocacy explicitly framed in terms of broader economic, social, and political rights as defined under international human rights law and covenants, including rights to self-determination by indigenous ethnic minorities. The Indigenous Community Support Organization (ICSO), for example, was created in 2007 to support ethnic minority communities to undertake advocacy for natural resource access rights. It also promotes civil society networking at the national level, advocating principles of the International Covenant on Economic, Social, and Cultural Rights, in particular to strengthen the development and implementation of legal protections for indigenous ethnic minority rights. The Land Law, Forest Law, and even the 1993 Constitution reference such rights, but only in a brief and general way, lacking the detail required to provide more specific protection (Sao Vansey, 4 May 2009, interview).

One benefit of national advocacy and international networking is increased attention by UN agencies and international development organizations operating in Cambodia to ethnic minority rights, says Srey Sras Panha, facilitator of the Indigenous Rights Active Members network (IRAM). This attention has opened up opportunities to influence donor programming. CLEC, in addition to its support for local-level resolution of land conflicts, as described above, has recently expanded its approach to include direct dialogue with donor agencies. Targeting development programs in the natural resource sectors that influence the actions of government and private investors, their aim is to protect the rights of and increase the benefits for local people (Kit Touch, 4 May 2009, interview).

Many activists, however, describe tensions between grassroots networks and local community leaders, on the one hand, aiming to address specific urgent conflicts, and national advocacy groups concerned with preserving their broader influence on government policy and implementation. Acknowledging this tension, Srey Sras Panha of IRAM explains: "Advocacy is not just about confrontation, it's about supporting the government so that they think about the benefits for the whole society. We don't criticize those who adopt direct confrontation. All these approaches are needed."

The different political allegiances among leaders of various human rights organizations also limit their combined effectiveness. During recent years, a number of distinct umbrella groups have emerged. According to Kit Touch, future progress within the human rights community depends on bridging these divides across organizations and across levels of civil society activism:

> *[To more effectively pressure the government], there must be a voice that really emerges from villagers to be heard, which means civil society organizations supporting community groups, and civil society organizations cooperating well together. Then the community voices will be strong and civil society organizations' voices will be strong, and the joining of these two voices will bring change to Cambodia. I want to see this happen so that the change comes.*

Building Coalitions to Secure Resource Rights

Rights-based approaches are evolving; yet, most forest-dependent communities in Cambodia are more vulnerable than several years ago, not less so. While there are local gains, many human rights activists feel a general sense of malaise at the lack of more systemic, tangible results. Environmental conservation and community development organizations are also making progress in many locales; but the broader trends of resource degradation and expropriation continue at an unsettling pace. What's needed?

Our analysis of the experience of civil society organizations indicates the most promising strategy is one that incorporates the following elements:

- learning from experience, with a self-critical examination of the advantages and limitations of the three approaches described above;
- linking more effectively across geographies and sectors, so that local rights advocacy influences national policy, and legal and institutional reform processes;
- building collective action to pursue resource management objectives at the ecosystem scale.

Such an integrated strategy necessarily entails aligning the actions of multiple stakeholders domestically and internationally.

Learning from experience

Each of the three approaches described in the previous section faces different limitations. Community mobilization in defense of land rights sometimes moves quickly to direct action without first exhausting other avenues, thus missing opportunities for negotiation. Many communities lack the resources and time to organize, or exposure to understand the implications of some kinds of development until it reaches a crisis, such as tractors clearing their land. Many are not aware of the avenues open to them to complain or negotiate, and some of those avenues may even be obscured or obstructed. Others simply have no faith that the legal and governance systems will work in a just way. Lastly, there is often little or no advance warning of company actions, particularly at community level, meaning that communities may have no choice but to act at the crisis point.

Nevertheless, moving to impromptu or poorly planned confrontations with companies is a risky venture and one that can be easily misrepresented as illegal or violent, even when it is not. The communities may lose the sympathy of potential allies, including donor agencies. Where communities take a more multilayered and proactive approach, which may include strategic and well-managed direct action, they have often been more successful at slowing land clearing or evictions and have sometimes even had successes. Such a strategy includes raising issues informally with local leaders; requesting documentation of approvals of company activities; filing complaints at various levels; seeking media coverage; conducting demonstrations or marches outside of the conflict area (such as in front of government offices); reporting problems to watchdog NGOs and requesting their investigation; documenting and circulating evidence of land and forest violations; and sometimes gaining the support of NGOs to assist with legal advocacy.

This does not preclude communities from using direct non-violent action. In fact, the use of a combination of strategies over time has the effect of strengthening community resolve, demonstrating solidarity, generating external support and sometimes more public sympathy, and thus gaining more leverage to pressure companies into some form of mediation or to convince authorities to take action. It may also yield more disciplined and strategic use of active non-violence when it is deemed necessary. However, this kind of approach requires appropriate community leadership, strong organizational skills, strategic and analytical thinking, and some level of knowledge and/or experience to inform analysis and planning – all areas that remain underdeveloped in Cambodian communities.

A focus on community forestry as a vehicle to secure local resource rights faces different constraints – namely, the narrowly defined steps and procedures proscribed by government regulation. NGOs adopting the approach risk losing sight of more significant threats to local food security and opportunities to support the rights that forest communities are demanding. A sectoral project-based approach also sometimes obstructs effective advocacy across a range of resource and rights issues. Negotiations aimed at achieving project objectives may restrict the range of concerns that communities might otherwise seek to raise, and in some instances communities may even be cautioned against raising complaints, as this could reflect poorly on the supporting NGO, government agencies, or local leaders. As a result, some grassroots advocates have come to the opinion that community forestry is at least as likely to divest people of land and resources as it is to help them protect claims because agreement to a defined community forest area may be perceived as tacit acceptance of concessions or other types of forest conversion across the larger landscape. When community forests are small or degraded and their capacity to control access is weak, villagers also risk having their tenure rights revoked years later on grounds of poor management. Many are increasingly suspicious of the NGOs and government agencies that try to persuade them into such agreements. Indeed, conservation NGOs are typically bound by memoranda of understanding with government which stipulate that they may not advocate on conservation issues. Therefore, there is little information available to the public about the cumulative trends regarding environmental degradation, resource exploitation, and the threats that these pose to local livelihoods in Cambodia. This contributes to a citizenry who are not broadly engaged on environmental issues despite the consequences that they may bear later.

Done in appropriate ways, broader rights advocacy could potentially motivate and inform a more sustained movement for social justice. However, the success of this depends on how it is undertaken. A common practice of NGOs is to train communities about their rights in ways that are too general, without relating those rights to people's own situation or avenues for redress, and too technical or legalistic, without providing opportunities for people to reflect and analyze for themselves. Most successful applications of a broader rights approach in Cambodia, by contrast, have included foundational activities (such as community media) that value and reinforce community identity, combined with consultative processes that engage communities in learning about and exploring their rights, and analyzing their own situations in relation to these rights. Sometimes these efforts are linked up at national level. For example, during 2008 to 2009, IRAM and the East–West Management Institute (EWMI) supported community representatives from around the country, and represented various ethnic and special interest groups in order to reflect on their economic, social and cultural rights. The results of these consultations helped to inform the writing of a civil society "shadow report" on Cambodia's status with regard to the International Covenant on Economic, Social, and Cultural Rights (Indigenous Community Support Organization et al, 2009). Perhaps more importantly, in the process of consultation, diverse communities identified common interests and made the decision to cooperate together to coordinate complaints about concessions and other land conflicts at the national level.

Linking across geographies and sectors

A major obstacle today is that the structures of decentralized governance in rural Cambodia often shrink away just when they are most needed – at the point of conflict. Grassroots advocates report that commune councils in some locales have simply stopped meeting once communities have urgent demands. In other places, commune councilors or village chiefs have had their participation in local governance limited after taking a strong stand in defense of community resource rights. There have even been cases where local authorities backing up community claims are charged, convicted, and imprisoned on criminal charges such as "falsifying information" (CHRAC, 2009). With significant investment going into building the capacity of these institutions to facilitate local development planning at commune, district, and provincial levels, there is an opportunity to strengthen their role in anticipating and resolving resource disputes. More broadly, there is an opportunity to increase the effectiveness of initiatives aimed at improving governance and social accountability if these are linked more systematically to addressing trends in environmental resource allocation and use.

Similarly, most human rights organizations still give little emphasis to natural resource livelihood issues until they are expressed as direct conflicts, particularly involving violence or loss of property. Thus, eviction of urban squatter communities, for example, receives more attention than does encroachment on forested areas essential to rural communities. Human rights organizations in Cambodia also typically have not advocated against loss of key environmental resources because they see this as a purely environmental issue, rather than a matter of fundamental rights to food, water, and livelihood.

If joined together, constituency-based groups and organizations focused on human rights, community development and rural livelihoods, environmental management, decentralized governance, and public-sector reform could make a much stronger case for the role of human rights and environmental governance as central to the prospects for national development, not only as matters of concern to particular marginalized segments of society. Current trends with natural resource allocation and resource degradation, displacing poor people and their source of livelihoods, represent a serious threat to social stability. Conversely, if managed in ways that support social equity and long-term sustainability, Cambodia's environmental resources still offer the opportunity to underpin broad-based community-driven development.

Making these links in a sustained way need not depend on the initiative of individual community development organizers, rights advocates, or grassroots leaders alone; it also requires a reorientation of development assistance. Most development agencies still take a largely technical approach, providing expert advice on legal development, capacity-building, and technical advice for sectoral management, without explicitly addressing the social and political dimensions of the problems that they aim to resolve. As Zimmerman and Kruk (2003) argued in commenting on land reform in Cambodia, "policy-oriented development cooperation fostering good governance must also be ready to tackle the issues of corruption and deal with both open and concealed power structures".

Building collective action at the ecosystem scale

At present, community-based natural resource management in Cambodia is primarily site specific, whereas the trends affecting community livelihoods play out at broader ecosystem scales. Defending Cambodia's environmental sustainability, while also protecting land and natural resource rights, requires new paradigms for development, as well as longer-term vision. It requires moving beyond advocacy mainly *against* rights violations to also advocating *for* innovative solutions that appeal to decision-makers, particularly by demonstrating how strengthening local rights can also contribute to sustaining resources of economic or social value for others beyond local communities.

An example is the work that EWMI and others are supporting at local, provincial, and national levels in an effort to preserve Prey Lang, the largest lowland evergreen forest in mainland Southeast Asia that remains unprotected. EWMI is advocating a co-management model for Prey Lang, such as a conservation concession that would enable communities to be central to management and protection of Prey Lang as an intact ecosystem, including protecting the watershed functions that it provides to downstream areas, with technical support from government and NGOs, and revenues shared at various levels. Recently, the Forestry Administration has responded with interest to public invitations by the Prey Lang Network, a coalition of local NGOs and community groups, to partner on forest management in the area.

At present, however, no legal framework exists to support such ecosystem-wide community-based forest management alternatives. The legal definition of community forestry in Cambodia restricts community forestry areas to production forests under the authority of the Forestry Administration. A broader range of options is now being explored, including community conservation forestry (in official protected areas), partnership forestry (where responsibility rests with commune councils who also control revenue), and community-based production forestry (in floodplain forest zones). The National Community Forestry Program, currently under development, has an objective to review and revise the legal framework for community forestry so that such alternative modalities are included (Bampton, May 2009, and Phan, July 2010, interviews).

Yet, even if the legal space for such alternatives is created, and, indeed, beforehand, it is necessary to build effective networks to advocate for and to implement resource protection at the broader ecosystem scale. The communities who reside within and around such potential co-management conservation zones are typically dispersed, poor, and politically marginalized, and therefore face severe limits as to what they can achieve on their own. The goal of groups assisting them, therefore, should not only be to ensure that their voices are heard in defending against rights violations. Civil society groups and development agencies need to invest more proactively in building the influence and the capacity of broader networks of communities to negotiate with local and provincial governments, and to partner among themselves in order to be able to assume joint responsibility in ecosystem-level co-management, and navigate the range of challenges that they will inevitably face. This requires an effective coalition of organizations to support such efforts – spanning the fields of human rights, law, policy, environment, community development, and governance.

Conclusions

What lessons does the last decade of experience aimed at securing rights for forest-dependent communities in Cambodia offer for those working on similar challenges elsewhere? For the human rights community, this experience reaffirms the need to define rights in a broad perspective that incorporates access to environmental resources for rural communities as a foundation for economic and livelihood security, linkages that are especially important in the context of post-conflict development assistance and peace-building (Ratner, 2011). It also encourages reflection aimed at understanding the work of human rights advocacy NGOs within a social context, asking how their choices of priority issues for attention, the alliances they build, and their processes of working contribute to problems or solutions for local communities and facilitate community mobilization in a climate of scarce resources.

For the international environmental conservation community, this experience illustrates the need to relate resource conservation and environmental sustainability to human rights, economic development, and democratization (Ribot, 2007; Sunderlin et al, 2008). It underscores the importance of policy advocacy and civil society mobilization to improve downward and horizontal accountability of decision-makers in the governance of forest resources (Agrawal et al, 2008). Perhaps even more importantly, it signals the critical role of building supportive relationships among organizations working with resource-dependent communities from human rights and development perspectives, including approaches that engage communities as partners and co-managers of environmental resources.

For the international development community, the experience in Cambodia highlights the limitations of short-term project-based approaches, and, in particular, the limitations of technical responses to solve deeply embedded political and social problems. It points to the need for a longer-term, more integrated strategic look at development trends and how to influence these realistically, recognizing the opportunities for sectoral policy and institutional reforms to influence broader governance arrangements (Brinkerhoff, 2000; DFID, 2010). It means investing in the development of civil society beyond formal NGOs, extending to grassroots social justice networks – recognizing the particular challenges that these groups face in addressing immediate subsistence and livelihood needs, while organizing to address broader threats. It means supporting links among civil society groups across international borders in order to address transboundary risks and opportunities (Sneddon and Fox, 2007). And this inevitably requires more strategic coordination, not only regarding sectoral investments, but also regarding the collective influence of development partners on policy and institutional change in practice.

Acknowledgements

Sincere thanks to the following for interviews conducted during April to May 2009, and July 2010, and for providing additional information: Amanda Bradley (PACT); James Bampton (RECOFTC); Phan Kamnap (Forestry Administration); Prak Marina (Siem Reap Provincial Forestry Department); Pen Bonnar (ADHOC); Sao Vansey (ICSO); Kit Touch (CLEC); Srey Sras Panha (NGO Forum – IRAM); Yim Leang Y (CCD); Suon Sareth (CHRAC); and Pyrou Chung (EWMI – PRAJ). Several local activists interviewed have requested to remain anonymous because of sensitivities with their work. Thanks as well to Mith Samonn and Keat Thida (WorldFish Center) for providing research assistance. Interviews quoted in the text were conducted in Khmer (apart from Bradley and Bampton) and translated by B. D. Ratner. An earlier version of this chapter was presented at the international workshop Towards a Rights-Based Agenda in International Forestry? at the University of California, Berkeley, on 30 to 31 May 2009. We wish to thank the organizers, Thomas Sikor and Johannes Stahl, as well as workshop participants, particularly Peter Leigh Taylor and William Sunderlin, for fruitful discussions. Thanks as well to Jeremy Ironside, who provided helpful comments on a draft. Any errors are the responsibility of the authors. The views expressed in this chapter are those of the authors, not the institutions that they represent.

References

Agrawal, A., A. Chhatre, and R. Hardin (2008) "Changing governance of the world's forests", *Science*, vol 320, no 5882, pp1460–1462

Bampton, J., H. Chandet, and S. Brofeldt (2008) "Conflict relating to community forestry in Cambodia: A case from Kbal Damrei Commune, Kratie Province", Paper presented at the International Conference on Tropical Forestry in a Changing World, Kasetsart University, Bangkok, Thailand, 17–20 November

Brinkerhoff, D. (2000) "Democratic governance and sectoral policy reform: Tracing linkages and exploring synergies", *World Development*, vol 28, no 4, pp601–615

CHRAC (2009) *Losing Ground: Forced Evictions and Intimidation in Cambodia*, Cambodia Human Rights Action Coalition, Phnom Penh

Davis, M. (2005) "Forests and conflict in Cambodia", *International Forestry Review*, vol 7, no 2, pp161–164

DFID (UK Department for International Development) (2010) *The Politics of Poverty: Elites, Citizens, and States*, DFID, London

Indigenous Community Support Organization, NGO Forum on Cambodia, Asian Indigenous Peoples' Pact Foundation, and Forest Peoples' Programme (2009) *The Rights of Indigenous People's in Cambodia*, Submitted to the UN Committee on Economic, Social, and Cultural Rights (42nd Session, April 2009), www.ciddhu.uqam.ca/documents/Annexes_Cambodia.pdf

IPNN (Indigenous People NGO Network) (2010) *The Rights of Indigenous Peoples in Cambodia*, Report submitted to the UN Committee on the Elimination of Racial Discrimination (76th session), IPNN, www.elaw.org/node/5349

Le Billon, P. (2000) "The political ecology of transition in Cambodia 1989–1999: War, peace and forest exploitation", *Development and Change*, vol 31, no 4, pp785–805

Le Billon, P. and S. Springer (2007) "Between war and peace: Violence and accommodation in the Cambodian logging sector", in W. De Jong, D. Donovan, and K. Abe (eds) *Extreme Conflict and Tropical Forests*, Springer, Amsterdam, the Netherlands, pp17–36

Ratner, B. D. (2011) "Building resilience in rural livelihood systems as an investment in conflict prevention", in H. Young and L. Goldman (eds) *Livelihoods, Natural Resources, and Post-Conflict Peacebuilding*, Earthscan, London

Ribot, J. C. (2007) "Representation, citizenship and the public domain in democratic decentralization", *Development*, vol 50, no 1, pp43–49

Shadravan, F. (2003) *Five Days in November*, Documentary film, A Mind's Eye Production, RECOFTC and Community-Based Natural Resources Management Learning Institute, Phnom Penh

Sneddon, C. and C. Fox (2007) "Power, development, and institutional change: Participatory governance in the Lower Mekong Basin", *World Development*, vol 35, no 12, pp2161–2181

Sunderlin, W. D. (2005) "Poverty alleviation through community forestry in Cambodia, Laos, and Vietnam: An assessment of the potential", *Forest Policy and Economics*, vol 8, no 2006, pp386–396

Sunderlin, W. D., J. Hatcher, and M. Liddle (2008) *From Exclusion to Ownership? Challenges and Opportunities in Advancing Forest Tenure Reform*, Rights and Resources Initiative, Washington, DC

Talbott, K. (1998) "Logging in Cambodia: Politics and plunder", in F. Z. Brown and D. G. Timberman (eds) *Cambodia and the International Community*, Asia Society, New York, pp149–168

World Bank, UNDP, and FAO (1996) *Cambodia: Forest Policy Assessment*, World Bank, Washington, DC

Zimmerman, W. and G. Kruk (2003) "Land reform in South-East Asian transition countries: Cambodia's path as a post-conflict country in transition", *Agriculture and Rural Development*, vol 10, no 2, pp54–57

15

Forest-Based Social Movements in Latin America

Peter Cronkleton and Peter Leigh Taylor

Throughout Latin America, substantial areas of forestland have been devolved to rural indigenous and traditionally forest-dependent rural populations (Barry and Taylor, 2008; Larson et al, 2010; see also Chapter 2 in this volume). Frequently forest-based social movements have played important roles in pressuring national governments to demarcate and recognize property rights (Brechin et al, 2003; Merry et al, 2006; Cronkleton et al, 2008; Sunderlin et al, 2008). These groups have relied on collective action to assert claims over forest resources, defend productive interests and traditional livelihoods, and give voice to rural views of development policy. Where they have emerged successfully, the rights gained by forest-based social movements have dramatically improved the property rights security of participants and expanded control over immense territories, usually with collective rights over communal lands. These movements often seek recognition of customary or traditional rights staked out historically and try to ensure that local people maintain control over forest benefits. However, the property and resource rights acquired by such movements also entailed significant limitations that can undercut overall benefits (Larson et al, 2008; Sunderlin et al, 2008). In addition, the incremental advance in rights recognition can defuse the original catalyst for collective action. As a result, leaders of grassroots movements need to work to maintain relevance and defend newly gained rights (Cronkleton et al, 2008).

Sikor and Stahl's introduction to this book (see Chapter 1) raises the question of what strategies are suitable for pursuing recognition of community forest rights. In order to examine this and related issues, this chapter draws on cases from Brazil, Guatemala, and Bolivia to identify factors in the emergence of forest-based social movements, analyzes characteristics of the rights they have achieved, and discusses the continued challenges that they face. The social movements examined include the Rubber Tappers' Movement in the Brazilian Amazon (CNS), the Association of Forest Communities of the Guatemalan Petén (ACOFOP), and the Confederation of Indigenous People of Eastern Bolivia (CIDOB).

Expansion of Forest Properties and the Role of Collective Action by Rural Social Movements

Over the past several decades, there has been a dramatic shift in forest property rights to indigenous and traditionally forest-dependent people in Latin America. In the region, tens of millions of people occupy forestlands (Kaimowitz, 2003), claiming either customary rights or having achieved formal recognition of their rights to use and live in the forest. Forest tenure reform has recognized local rights over approximately 270 million hectares, or about 25 percent of the subtropical and tropical forestlands in Latin America (Chomitz, 2007; also see Sunderlin et al, 2008, and Chapter 2 in this book). Civil society and activist grassroots organizations have played significant roles in the process of rights transfer throughout the region. Forest-based social movements have used a variety of approaches to exert influence over this process. Significantly, in each case, success in organizing has been achieved largely by creating regional organizations capable of linking into national and international stakeholder networks (Stewart, 2006; Cronkleton et al, 2008; Borras, 2010; see also Chapter 13 in this volume on the need to organize across scales, and Chapter 14 on the influence of external actors).

At first glance the forests of Latin America would not seem to be an ideal context for social movements to develop. The region's forest-dependent rural people often are dispersed, isolated, and economically and politically marginalized. However, at times, these people coalesce into organized groups and networks that are able to push forth a common agenda, defend interests, and pressure governments to recognize and respect their rights. In this chapter, we examine in detail these factors through Brazilian, Guatemalan, and Bolivian cases. As discussed below in detail, the reasons for these movements' success in organizing lie in complex historical processes shaped by a variety of factors, including the presence of charismatic leadership, the capacity to achieve consensus on the nature of external threats, and the manner to respond, as well as the existence of political conditions making crucial alliances with external state and private actors possible (Cronkleton et al, 2008).

In this chapter we will examine prominent cases where forest-based social movements have arisen and eventually achieved strengthened property rights, particularly through innovative forest tenure models that supported forest livelihoods. The three cases include the Rubber Tappers' Movement in Brazil, the ACOFOP community forest organization in Guatemala, and the CIDOB indigenous federation in Bolivia. It is important to note that although these movements were made up of forest-dependent people, they were largely motivated by livelihood and development issues, which has at times created tensions with their environmentalist and conservationist allies.

Rubber tappers and extractive reserves in Brazil

The Rubber Tappers' Movement in the Brazilian Amazon is an iconic example of grassroots activism to defend forest livelihoods. Starting in the 1970s as a movement of impoverished forest people confronting development change that threatened their livelihoods, over the following decade the rubber tappers adopted a strategy of collective protest, formed alliances with international environmental and human rights organizations, and eventually influenced Brazilian conservation, development, and tenure policies at the regional scale. The movement's efforts led to the formal recognition of property rights for forest communities through the creation of extractive reserves and agro-extractive settlements. These new property types dramatically expanded areas of forestland controlled or owned by community-level actors in the Brazilian Amazon.

Rubber tappers' livelihoods are heavily dependent upon the extraction of non-timber forest products (NTFPs). Their origin can be traced back to the start of the rubber boom during the late 19th century when large estates (*seringals*) divided up the region's forests and imported labor to gather latex from wild stands of rubber trees (*Hevea brasiliensi*). When rubber prices dropped in the early 20th century, the estate system collapsed and the laborers were left on their own to develop agro-extractive livelihoods based on the extraction of rubber and Brazil nuts (*Bertholletia excelsa*) combined with subsistence agriculture. This was a relatively benign forest use, but required extensive areas of intact primary forests for families to make a living. Many families became semi-autonomous producers, which allowed them to continue to occupy the forest and eventually claim usufruct rights to property.

During the 1970s, Brazil's military government adopted new policies to promote Amazon development to replace the pre-existing forest extractive economy. The government provided incentives for the entrance of large-scale enterprises and established huge colonization projects to resettle displaced and landless families from other regions (see Schmink and Wood, 1992). Resistance to official development policy grew among rubber tappers in the countryside as deforestation increased. In the process, the state refused to recognize the traditional land rights of the rubber tappers occupying the forests, and instead offered to relocate them to colonization projects. However, the colonization plots offered were too small for their traditional extractive activities, and most relocated families lacked capital to invest in agriculture. Faced with extreme hardship, many families abandoned their homesteads and moved to peri-urban shanty towns (Bakx, 1988). News spread of the difficulties faced by families pushed into colonization projects, further fueling rural unrest.

During this time the potential for violent conflict increased. Catholic lay volunteers and union organizers began forming community organizations, explaining legal rights and discussing the collective interests of isolated forest families. As local leaders emerged, they made explicit efforts to raise the consciousness and build consensus around the rubber tappers' collective interests – linking rural well-being to the forest became a key component of an emerging rubber tapper identity (Grzybowski, 1989).

This led, in 1985, to the first national meeting of rubber tappers and the formation of the National Rubber Tappers' Council (*Conselho Nacional dos Seringueiros*, or CNS). This was the first national organization capable of effectively representing the interests of rubber tappers and other forest workers in Amazonia (Schmink and Wood, 1992). One of the innovative ways in which the Rubber Tappers' Movement developed to resist expulsions was through the use of a non-violent collective action known as "*empate*" in which families from surrounding communities occupied forest areas slated for clearing by ranchers. While unable to halt deforestation completely, this non-violent activism drew national and international attention to the rubber tapper cause. The CNS rapidly began to garner greater attention through its collective actions and strategic alliances with international environmental organizations, notably the Environmental Defense Fund and the National Wildlife Federation (Schmink and Wood, 1992). These contacts allowed the rubber tappers to communicate with international development banks and resulted in a re-evaluation of support for Brazilian development policy for the Amazon (Schwartzman, 1991).

Eventually, the federal government gave in to the CNS's principal demand for the recognition of traditional property rights by creating agro-extractive settlements (known as PAEs for the Portuguese acronym) in 1987 and extractive reserves (RESEX) during 1990. PAEs are a property type that recognized the rights of existing occupants of forest areas by providing communal title and allowing them to maintain their settlement patterns and customary extractive livelihoods. Extractive reserves are a type of conservation area that recognized access and management rights of rubber tapper families, allowing residents to continue traditional activities. Neither model represented a redistribution or resettlement effort, but instead recognized the *de facto* rights of people already living in the forest. There are currently 48 RESEX in the Amazon covering a total of 12 million hectares (IBAMA, 2009).

While property rights were a main goal, the strength of the rubber tapper movement was crucial in supporting allies to gradually take control of local and state governments. In the state of Acre, where the Rubber Tappers' Movement was born, the movement aligned with similar groups and like-minded political parties to systematically take control of numerous municipalities and eventually, in 1998, the state government. Their government allies prioritized the maintenance of the state's forests and began implementing policies and strategies to promote sustainable development in partnership with the state's traditional forest people (Kainer et al, 2003), pushing a pro-forest, pro-community agenda to the forefront of regional debates.

However, there have also been drawbacks for the movement from this success. Much of the leadership has been pulled into official office due to electoral success, and the activism of the base has dissipated. One consequence has been weaker participation in negotiating environmental regulations affecting forest use by smallholders and communities. The RESEX provide secure property, but the strict conservation rules are inflexible and complicate efforts by residents to diversify or adapt their production. In some cases, rules could be modified with the adoption of RESEX-wide management plans; but because these have to be devised by RESEX councils composed not only of residents, but also of government agencies and other stakeholders, they have remained a bottleneck constraining change.

Forest people and community concessions in Guatemala

The Association of Forest Communities of the Guatemalan Petén (ACOFOP) is a grassroots forest organization that emerged because of the threats to communities' forest access and well-being, posed by the initial design of the Maya Biosphere Reserve (MBR). Currently, ACOFOP consists of 23 member communities and organizations representing some 14,000 individuals in 30 communities who successfully adopted a collective strategy to gain timber concessions in the area around the MBR. Its members now manage approximately 500,000ha of some of the best managed forests of the Petén, nearly 70 percent of which has been certified as sustainably managed by the Forest Stewardship Council (ACOFOP, 2005). By 2010, the community concessions had succeeded in generating significant positive social, economic, and conservation impacts upon the MBR's multiple uses zone (Nittler and Tschinkel, 2005; Bray et al, 2008; Monterroso and Barry, 2009; Taylor, 2010), although more systematic study of their impact is needed.

ACOFOP emerged out of a context of poor governance and conflict. The Petén was one of Guatemala's most geographically and politically isolated regions and one that has been shaped by competition among diverse stakeholders to control its wealth of natural resources. A major group of residents includes descendants of migrants in extractive communities who appeared during the 1920s to exploit natural stands of important NTFPs (xate and chicle). More recent colonist migrants arrived to invest in agriculture, and some expanded into informal logging enterprises. During recent decades, these smallholders were joined by waves of larger agro-industries, ranchers, and oil companies that entered the region, and greatly intensified conflicts (Gómez and Méndez, 2005; Monterroso, 2007).

With increased deforestation, fire, and the looting of archaeological heritage, the Guatemalan government took action in 1990 by creating the Maya Biosphere Reserve. The original policies for the multiple use and buffer zones of the MBR severely restricted resource access by the local population, triggering serious conflicts. By the late 1990s, it had become clear that the government was experiencing great difficulty in managing the contradictions in state policy and social conflict arising from the MBR. As a result, there was growing interest in adjusting policies. Moreover, the MBR represented a key piece of a larger international conservation strategy in Central America; therefore, influential international stakeholders, including donor governments and international conservation organizations, wished to diffuse governance conflict surrounding the reserve.

One proposal discussed was the allocation of timber concessions around the MBR. Concessions were seen as a promising option for mitigating social conflict and promoting broader community involvement in the reserve, and, as a result, facilitating more effective management. However, an intense debate ensued over whether to grant forest management concessions to the timber industry or to communities. In the midst of this debate, community-based groups began organizing to lobby for forest management rights. Community groups, including both those dependent upon NTFP extraction, as well as others involved in informal timber, united as an organization that became ACOFOP. The controversy was ultimately resolved by the granting of forest management concessions to 12 local communities and 2 local forest industries (Nittler and Tschinkel, 2005). The collective influence of the communities' own local association was crucial to the communities' winning the concessions. However, international support for granting community concessions probably signaled more an opposition to industrial concessions than a strong belief in local communities' capacities (Gómez and Méndez, 2005; Cronkleton et al, 2008). The concessions offered 25-year management rights with the possibility of renewal; but the rights were contingent upon the preparation of a sustainable management plan and the achievement of Forest Stewardship Council (FSC) certification.

Gaining access to the concessions proved to be only an initial catalyst for collective action by ACOFOP. The community concession agreement originally required communities to sign exclusive technical assistance contracts with local non-governmental organizations (NGOs), themselves linked to international conservation organizations. The relationships between communities and NGOs were not smooth. ACOFOP member communities needed external assistance to comply with the rules (i.e. preparing management plans and attaining certification), but wanted it on their own terms. ACOFOP and its associated communities and organizations complained that the NGOs' methodologies were "top down" and did not allow for the development of local capabilities for integrated forest management, administration, and business. With growing tensions between NGOs and communities, the concession communities countered with pressure from ACOFOP and support from international allies. Local resistance eventually brought an end to the obligatory NGO technical support model in 2001. Communities gained greater freedom to seek technical assistance adequate for their needs. Several NGOs continue today to provide technical assistance, though on terms more acceptable to the communities (see Cronkleton et al, 2008; Taylor et al, 2008).

Today, the community concessions of the Petén face significant obstacles. Some member organizations struggle, especially those that are institutionally weak, were granted the least commercially valuable forests, causing some families to lose interest, and that have their rights contested by powerful private landowners present in their areas (Nittler and Tschinkel, 2005; Monterroso, 2007). ACOFOP also faces skepticism and outright opposition from industry and some NGOs (Gómez and Méndez, 2005; Monterroso, 2007). Today, ACOFOP and its members seek to develop a more integrated management approach placing more emphasis on NTFPs to help consolidate the community concessions and respond to conservationist concerns about timber exploitation (Taylor, 2010).

In addition, the right of communities to participate in forest management continues to be an ongoing struggle. ACOFOP will need to lobby to extend concession contracts after the initial 25-year period; but, more importantly, will need to remain vigilant for political shifts that could undercut the concession arrangements. For example, in 2003 a new proposal to develop a protected area in northern Petén called the Mirador Basin was introduced. The Mirador Basin project proposal was a major effort to expand strict conservation areas protecting the region's forests and archaeological heritage. The proposal would have halted the forest management and livelihood strategies in concessionaire communities, replacing them with private "sustainable eco-tourism initiatives" with monetary compensation to halt timber extraction (Monterroso, 2007; Monterroso and Barry, 2009).

The Mirador Basin project was successfully overturned by Guatemala's Supreme Court in 2005 after action brought by ACOFOP on behalf of the community concessions and their allies (including some industrial groups) (ACOFOP, 2005). The future of the Mirador Basin and its relationship to community management of the multiple-use buffer zone continues to be unclear, and new proposals to change the classification of the area have continued to appear (Cortave, pers comm, 2010; Schmidt, 2010). ACOFOP and its associated communities have stated that they do not oppose development, but rather wish to have community interests represented in planning processes. By 2010, ACOFOP and its associated communities had succeeded in gaining more formal participation in negotiations related to these and other proposals affecting the community concessions (Taylor, 2010).

Indigenous struggle for forest properties in Bolivia

Indigenous social movements have played an important role in defining forest property rights in the forests of lowland Bolivia. Social movements involving lowland indigenous peoples coalesced around the issue of property rights during the last decades of the 20th century, and eventually pressure from indigenous grassroots organizations resulted in the creation of a new type of property called a *tierra communitaria de origen* (TCO). The concepts underlying TCOs are to recognize the ancestral rights claims and provide sufficient land and resources for maintaining traditional livelihood through the issue of communal property titles. Internal organization and resource use would be based on traditional practice and custom (*usos y custumbres*) or could include commercial forest use with an approved management plan. Territories tend to be immense multi-community and, at times, multi-ethnic landscapes. As a result of the inclusion of TCOs, the state has recognized indigenous access rights to forests for subsistence and commercial ends, and millions of hectares of lowland forests have been nominally titled in favor of indigenous peoples.

Prior to the 1990s, Bolivia's indigenous population in the lowlands had long been marginalized in remote inaccessible territory. Agrarian reform had secularized large mission estates, but indigenous peoples continued to lack recognized property rights and were dependent upon the region's non-indigenous elite, who held economic and political power (Jones, 1997). Although property rights for indigenous and other traditional rural peoples were weak or non-existent, they occupied remote forestlands that initially were seen as economically marginal, and the low population density allowed them to enjoy usufruct rights to forests and other resources.

However, starting in the 1960s and accelerating in the 1980s, indigenous peoples came under increasing pressure as initiatives to develop the eastern lowlands region brought road construction and development policy supporting the expansion of timber and agricultural frontiers, as well as petroleum exploration (Jones, 1997; Colchester et al, 2001). These changes increased pressure on resources, and indigenous rights to forests were contested. Indigenous peoples' traditional territories were not respected, deforestation increased, and they received little benefit from the change. In addition, top-down initiatives, such as "debt for nature swaps" involving indigenous lands, increased rural tension (Stocks, 2005). In response, indigenous peoples began to organize representative organizations. Eventually, the organizations united into an indigenous lowland federation called the Confederation of Indigenous People of Eastern Bolivia (CIDOB).[1] CIDOB is Bolivia's principal organization of lowland indigenous peoples.

As they had little political recourse, the indigenous movement began to use a strategy of mass marches, the most famous a march on the national capital known as the March for Land and Dignity in 1990. The marches drew attention to their plight, disrupted transportation, and placed pressure on the government. Eventually, the government capitulated and created nine indigenous territories for approximately 2.9 million hectares through presidential decrees; however, over time there was little political will to implement and enforce the decrees (Stocks, 2005). So, after these TCOs were decreed, the indigenous movement had to maintain pressure to ensure that the government followed through with titling and later to assure that TCOs were codified as a new tenure reform law was drafted and then ratified in 1996.

The inclusion of TCOs in 1996 tenure reform was intended to prioritize indigenous interests (Kay and Urioste, 2005); however, in practice, the government's approach to titling TCOs has indicated a lack of political will to confront hard choices and opposition from politically and economically powerful stakeholders (Cronkleton et al, 2009).

Early on, critics questioned the slow progress in TCO titling due to the prioritization of third-party land claims and forest concession rights (CEJIS, 1999; Tamburini and Betancur, 2000), which also denied them uniform and contiguous territories. In 2000, there were additional protests, including a third march that was known as the March for Land, Territory, and Natural Resources that allied the indigenous peoples with peasants from northern Bolivia seeking property rights of the Brazil nut-rich forests that they occupied (see Cronkleton and Pacheco, 2010). These protests were successful in shifting specific government policy, but did not produce sustained efforts by the government to fully title indigenous lands. As a grassroots movement, Bolivia's lowland indigenous peoples face major challenges in maintaining cohesion due to their multi-ethnic composition, geographic dispersion of their territories, and the government's tendency to respond in a gradual and piecemeal fashion.

A good example of the process can be seen in the Guarayos TCO in the Santa Cruz Department (Cronkleton et al, 2009). The Guarayos people participated in the indigenous marches in 1990, formed a representative organization entitled the Central Organization of the Native Guarayos Peoples (COPNAG)[2] in 1992, and in 1996 participated in the second indigenous march on the eve of ratification of the tenure reform law. In 1999, the government determined that their territorial needs were 1,349,882ha; however, in the subsequent titling process the government prioritized remote uncontested areas for titling, setting up titling units that ignored patterns of traditional land-use and population distribution. This strategy allowed titling to progress rapidly, and by 2007 about 73 percent of the spatial needs for the Guaryaos people defined by the government back in 1999 were resolved; but most of these lands were far from where the Guarayo population actually lived. During the drawn-out process, many of the non-indigenous third parties occupying lands within the original TCO demand were able to consolidate their holdings.

The use of collective action by Bolivia's lowland indigenous population was intended to gain recognition of indigenous territory. Currently, 60 TCO demands in the forested lowlands have been accepted by the government for more than 17.5 million hectares. The indigenous movement was successful in pushing for the creation of TCOs; but few are fully titled. Once that battle was won, there was a continuous need to maintain pressure on policy-makers to actually implement rights; but over time, with incremental change, the movement's ability to consolidate property rights has dissipated. Ten years after the passage of the tenure reform law, only 30 TCOs are at least partially titled, half of those acknowledged in Bolivia's lowlands. The titled area accounts for only 49 percent of the 9.6 million hectares demanded by these TCOs (CIDOB/CPTI, 2008).

What Rights Are Gained by Forest Social Movements and What Are the Limitations?

As these cases illustrate, despite significant geographic and temporal differences, forest-based social movements often passed through similar stages. Each originated in a region with sporadic presence of state institutions and, consequently, weak governance mechanisms. Rural producers developed livelihood strategies closely linked to, or dependent upon, the access and use of forest resources. As frontier change intensified, competition for resources increased and, as a result, conflict and degradation exacerbated problems faced by forest people. Typically, governments responded with conservation or resettlement initiatives, including protected areas or reserves often established with little or no consultation with local communities. Such measures delegitimized, restricted, or even eliminated community resource access and directly affected production, income, and subsistence systems, provoking tension and resistance, and under some conditions provided powerful catalysts for unity. Where they could recognize common ground, community-level stakeholders began to defend against external threats to their resource base, and with certain types of assistance, collective movements began to coalesce. After increased collective action and pressures for policy change, authorities eventually ceded greater control and access to communities and their grassroots movements. Significantly, although the groups examined here won their battles for greater property rights through collective action, they subsequently faced new challenges to maintain relevance to their members and to capitalize on their successes.

The collective action strategies adopted by the grassroots groups discussed here operated at a number of levels. In some cases, they adopted direct action, such as physically blocking deforestation, as happened in Acre, or by refusing to participate in programs, as with ACOFOP refusing conditions for working with certain NGOs, or by organizing large-scale demonstrations (e.g. protests in Guatemala City against the Mirador Basin initiative organized by ACOFOP and its allies, and indigenous marches, as occurred in Bolivia). By acting together, these groups have managed to raise the public profile of their members, significantly neutralize economic and political disparity with competing stakeholders, and draw attention to their plight and the impacts upon the forests they occupied. At another level, their high-profile actions and broad base of local support have made them attractive partners to external stakeholders. These social movements have worked actively to present themselves as credible participants in natural resource management and have sought to form alliances with powerful external stakeholders. Their initial success drew attention to their causes, helped them to form alliances with national and international conservation groups, and eventually obligated governments to respond to their demands. These were all outcomes unlikely to be achieved by individuals or ineffectively organized groups.

It is possible to identify several salient themes in the emergence and operation of these forest-based social movements. These include the importance of understanding common interest/threats; the lack of alternative recourse to influence policy; the role of external actors; and the challenge of maintaining relevance and cohesion among the movement after initial success. These issues illustrate underlying processes responsible for the course taken by these forest social movements.

Common interest and threat

The emergence of forest social movements is usually tied closely to collective actions in response to some common interest or threat. In the cases explored here, catalysts were generated by increased competition for forests and the exclusion of forest-dependent people. These changes were occurring with official government sanction of the rights of new actors while denying recognition of traditional rights of those occupying and using the forest. At the same time, in all three cases, structural opportunities for mobilization were created by democratization reform trends, though these were less dramatic in the Bolivian case. Faced with the loss of livelihoods and home, affected people developed approaches to work together to resist policies. The recognition of a shared agenda can act as a catalyst to overcoming isolation and difficult communication among dispersed forest populations. Because forest-dependent people are often dispersed and isolated from one another, the recognition of collective interests and subsequent organization of actions takes extra effort. Single events can spontaneously provoke collective response; but sustained movements require deeper conceptual understanding of the importance of working together to provide greater strength and clear strategy for success.

Lack of other recourse to defend their interests

Forest-based social movements emerge where people have little other recourse to influence policy. Frequently, individuals living in forests and affected by development policy have little leverage as individuals and few mechanisms to exert influence within the political system. In many countries in the developing world, courts are weak, democratic institutions are unresponsive, and government agencies are influenced by other more powerful constituents. Marshaling influence through collective action allows the views of the poor or politically marginalized to be heard and to grab the attention of other actors. Of course, these types of strategies are only possible where there is a sufficiently open society and free press to allow such public manifestation of dissent. In non-democratic societies, the participants risk repressive measures by state or violence from the threatened elite. Nonetheless, the leaders and members of the groups described here often had to endure and overcome violent opposition, but were still successful.

Alliances and networking

Connections among forest people can take place spontaneously; but outside assistance is often important for facilitating the organizational process. This is apparent, for example, in the case of the Rubber Tappers' Movement where the church, unions, and NGOs actively encouraged collective consciousness and informed rubber tappers of their rights. External actors do play important roles both in facilitating the emergence of social movements by informing people of their rights and establishing networks that allow rural people to learn of the plights of others. The assistance of external organizations is often crucial for getting over initial obstacles to organization and attaining necessary capabilities. External actors also find common cause alliances with these grassroots groups once movements are operational.

Governments may be able to ignore local protests; but the international allies of forest social movements can have more influence on bilateral agencies (banks or aid agencies), or on public opinion, so governments pay attention. Ironically, while the role of external actors is important for their formation, the validity or authenticity of groups is often based on their grassroots origins.

Maintain relevance and cohesion after success

In the Brazilian and, to a lesser degree, in the Bolivian case, previously marginalized community movements subsequently found significant representation in mainstream political structures. This increased formal political influence, however, was accompanied by an urgent need to maintain the movements' political relevance for their members. Once these grassroots movements succeeded in gaining more secure resource access, their organizations were obligated to shift focus to better support their members' changing needs. In order to maintain relevance, these groups needed to shift their attention to helping members take advantage of policy gains and improve livelihood conditions by exploring alternatives for expanding or diversifying forest production. These shifts have taken varying forms for improving forest-related income, identifying new production strategies, improving marketing capacity, and developing greater and more complex involvement in resource management.

However, success can also dissipate interest in participation in movement or allow internal differences and individualist tendencies to re-emerge. The struggle to gain property rights is usually a prolonged struggle; but initial success can take away a main catalyst for collective action. All three social movements discussed here faced the challenge of maintaining strong grassroots support in the face of continuing pressure and threats to their legitimacy. For example, in Guatemala, ACOFOP has continued to struggle to fight off attempts to change concession policies and eventually extend them. In Bolivia, indigenous peoples have gained demarcation for TCOs that provides a basic level of security; but titling processes have remained incomplete.

Consolidation of communities' significant success in natural resource management requires them to develop greater capacities to manage the complex politics of their resource base. For example, ACOFOP continually seeks to balance conservation and livelihood goals. While conservation is an important component of these organizations' agendas, participating members expect their organizations to continue to generate concrete improvements in their livelihoods, often measured in terms of employment and income. At the same time, the external legitimacy of grassroots forestry organizations' management of forest resources greatly depends upon the communities' ability to demonstrate that they are sustainably managing and conserving the resources. Although state support for community-based resource management has increased substantially, grassroots community organizations must continually work to maintain and extend policy support for their role. Moreover, communities' credibility with conservationist stakeholders often requires countering competing stakeholder claims or criticisms, often based on overly narrow data and observations. Community groups need to develop their capacity to recount their own experiences more accurately and persuasively. This requires strengthening of local skills in the collection, analysis, and presentation of data related to their forests and their activities in them (see Taylor et al, 2011).

Conclusions

Forest-based social movements have had dramatic positive influence over the well-being of forest-dependent rural people by advancing their rights over forest properties and ensuring that members are guaranteed rights to access and use of forest resources. Territorial gains of forest people have been impressive, with huge expanses of land turned over as a direct result of these movements. The property rights granted have meant more secure livelihoods and stable occupation of land. The resulting properties do not generally represent the redistribution of resources, but instead the recognition of pre-existing traditional rights that have allowed residents to stay in forests. While the successes of forest-based social movements have been significant, they continue to struggle to maintain and strengthen the legitimacy of their members' resource rights in the eyes of powerful stakeholders in their regions' forests, and at the same time remain responsive to the changing needs of their community constituencies. Nevertheless, the dramatic changes brought about by these movements have achieved historic public recognition of communities' rights to rely on forest resources for subsistence and commercialization, while managing them sustainably for future generations.

Notes

1 *Confederación de Pueblos Indígenas de Bolivia.*
2 *La Central de Organizaciones de Pueblos Nativos Guarayos.*

References

ACOFOP (Asociación de Comunidades Forestales de Petén) (2005) *Representamos al Proyecto Forestal Comunitario más Grande del Mundo*, www.acofop.org, 4 February 2005

Bakx, K. (1988) "From proletarian to peasant: Rural transformation in the State of Acre, 1870–1986", *Journal of Development Studies*, vol 24, no 2, pp141–160
Barry, D. and P. L. Taylor (2008) *An Ear to the Ground: Tenure Changes and Challenges for Forest Communities in Latin America*, Rights and Resources Initiative and Center for International Forestry Research, Washington, DC
Borras, S. M. (2010) "The politics of transnational agrarian movements", *Development and Change*, vol 41, pp771–803
Bray, D. B., E. Duran, V. Hugo, A. Velazquez, R. Balas McNab, D. Barry, and J. Radachowsky (2008) "Tropical deforestation, community forests, and protected areas in the Maya Forest", *Ecology and Society*, vol 13, no 2, p56
Brechin, S. R., P. R. Wilshusen, and C. E. Benjamin (2003) "Crafting conservation globally and locally: Complex organizations and governance regimes", in R. Brechin, R. R. Wilshusen, C. L. Fortwangler, and P. C. West (eds) *Contested Nature: Promoting International Biodiversity Conservation with Social Justice in the Twenty-first Century*, State University of New York Press, Albany, NY, pp159–182
CEJIS (1999) "Titulación de territorios indígenas: Un balance a dos años de la promulgación de la Ley 'INRA'", *Revista de Debate Social y Jurídico*, no 6, Enero-Abril, CEJIS, Santa Cruz
Chomitz, K. (2007) *At Loggerheads? Agricultural Expansion, Poverty Reduction, and Environment in the Tropical Forests*, World Bank, Washington, DC
CIDOB/CPTI (2008) *10 anos San-TCO: La lucha por los derechos territoriales indígenas de tierras bajas Bolivia*, Confederacion de Pueblos Indígenas de Bolivia CIDOB, Centro de Planificacion Territorial Indigena, CPTI, Santa Cruz de la Sierra, Bolivia
Colchester, M., F. MacKay, T. Griffiths, and J. Nelson (2001) *A Survey of Indigenous Land Tenure: A Report for the Land Tenure Service of the Food and Agriculture Organization*, FAO, www.forestpeople.gn.apc.org
Cronkleton, P. and P. Pacheco (2010) "Changing policy trends in the emergence of Bolivia's Brazil nut sector", in S. Laird, R. McLain, and R. Wynberg (eds) *Non-Timber Forest Products Policy: Frameworks for the Management, Trade and Use of NTFPs*, CIFOR and Earthscan, Bogor and London
Cronkleton, P., P. L. Taylor, D. Barry, S. Stone-Jovicich, and M. Schmink (2008) *Environmental Governance and the Emergence of Forest-Based Social Movements*, Center for International Forestry Research (CIFOR), CIFOR Occasional Paper no 49, Bogor, Indonesia
Cronkleton, P., P. Pacheco, R. Ibarguen, and M. A. Albornoz (2009) *Reformas en la tenencia de la tierra y los bosques: La gestión comunal en las tierras bajas de Bolivia*, CIFOR/CEDLA, La Paz, Bolivia
Gómez, I. and V. E. Méndez (2005) *Analisis de Contexto: El Caso de la Asociacion de Comunidades Forestales de Peten (ACOFOP)*, PRISMA, San Salvador, El Salvador
Grzybowski, C. (1989) *O testamento do homen da floresta (Chico Mendes por ele mesmo)*, Fase, Rio de Janeiro
IBAMA (2009) *Lista das unidades de conservação federais*, www.ibama.gov.br, accessed 25 July 2010
Jones, J. C. (1997) "Development: Reflections from Bolivia", *Human Organization*, vol 56, no 1, pp111–119
Kaimowitz, D. (2003) "Pobreza y bosques en América Latina: Una agenda de acción", *Revista Forestal Centroamericana*, pp13–15

Kainer, K., M. Schmink, A. Leite, and M. Da Silva Fadell (2003) "Experiments in forest-based development in Western Amazônia", *Society and Natural Resources*, vol 16, no 10, pp1–18

Kay, C. and M. Urioste (2005) *Bolivia's Unfinished Agrarian Reform: Rural Poverty and Development Policies*, ISS/UNDP Land, Poverty and Public Action Policy Paper no 3, Institute of Social Studies (ISS)/UNDP, The Hague

Larson, A., D. Barry, P. Cronkleton, and P. Pacheco (2008) *Tenure Rights and Beyond: Community Access to Forest Resources in Latin America*, CIFOR Occasional Paper no 50, CIFOR, Bogor, Indonesia

Larson, A., D. Barry, G. Ram Dahal, and C. J. P. Colfer (eds) (2010) *Forests for People: Community Rights and Forest Tenure Reform*, Earthscan, London

Merry, F., G. Amacher, M. Macqeen, D. Santos Guimares, E. Lima, and D. Nepstad (2006) "Collective action without collective ownership: Community associations and logging on the Amazon frontier", *International Forestry Review*, vol 8, pp211–221

Monterroso, I. (2007) *Nuevas Tendencias y Procesos que Influyen en el Manejo Comunitario Forestal en la Zona de Usos Multiples Reserva de Biósfera Maya en Peten, Guatemala*, CIFOR/ACOFOP, Guatemala City, Guatemala

Monterroso, I. and D. Barry (2009) *Tenencia de la Tierra, Bosques y Medios de Vida en la Reserva de la Biosfera Maya en Guatemala: Sistema de Concesiones Forestales*, CIFOR, Rights and Resources, and FLASCO, Guatemala

Nittler, J. and H. Tschinkel (2005) *Manejo Comunitario del Bosque en la Reserva Maya de la Biosfera de Guatemala: Protección Mediante Ganancias*, University of Georgia, Watkinsville, GA

Schmidt, B. (2010) "Ranchers and drug barons threaten rain forest", *The New York Times*, 17 July, www.nytimes.com/2010/07/18/world/americas/18guatemala.html?_r=1&th&emc=th

Schmink, M. and C. H. Wood (1992) *Contested Frontiers in Amazonia*, Columbia University Press, New York, NY

Schwartzman, S. (1991) "Deforestation and popular resistance in Acre: From local social movements to global network", *Centennial Review*, vol 35, no 2, pp397–422

Stewart, J. (2006) "When local troubles become transnational: The transformation of a Guatemalan indigenous rights movement", in H. Johnston and P. Almeida (eds) *Latin American Social Movements: Globalization, Democratization, and Transnational Networks*, Rowman & Littlefield Publishers, Lanham, pp197–214

Stocks, A. (2005) "Too much for too few: Problems of indigenous land rights in Latin America", *Annual Review of Anthropology*, vol 34, pp85–104

Sunderlin, W. D., J. Hatcher, and M. Liddle (2008) *From Exclusion to Ownership? Challenges and Opportunities in Advancing Forest Tenure Reform*, Rights and Resources Initiative, Washington, DC

Tamburini, L. and A. C. Betancur (2000) "Nuevo régimen forestal y territorios indígenas en Bolivia", in J. A. Martínez (eds) *Atlas Territorios Indígenas en Bolivia: Situación de las Tierras Comunitarias de Origen (TCOs) y Proceso de Titulación*, CPTI, Santa Cruz, pp213–228

Taylor, P. L. (2010) "Conservation, community, and culture? New organizational challenges of community forest concessions in the Maya Biosphere Reserve of Guatemala", *The Journal of Rural Studies*, vol 26, no 2, pp173–184

Taylor, P. L., P. Cronkleton, D. Barry, S. Stone-Jovicich, and M. Schmink (2008) *"If You Saw It With My Eyes": Collaborative Research and Assistance with Central American Forest Steward Communities*, Center for International Forestry Research (CIFOR), Forests and Governance Programme Series no 14, Bogor, Indonesia

Taylor, P. L., P. Cronkleton, and D. Barry (2011) "Learning in the field: Using community self studies to strengthen forest based social movements", *Sustainable Development*, doi:10.1002/sd.498

Part VI

Epilogue

16

A Way Forward: Forest Rights in Times of REDD+

Thomas Sikor and Johannes Stahl

REDD+ (reducing emissions from deforestation and degradation and enhancing carbon stocks) raises significant challenges to the recognition of forest people's rights. The challenges neither pose a uniform threat to forest people's rights, nor merely afford new opportunities for rights recognition. The chapters in this book addressing REDD+ reflect this openness, even though most authors express skepticism. In Chapter 2, Sunderlin notes that REDD+ may work in support of the transfer of tenure rights to forest people, or may expose people to new competition for the rights they already hold. In Chapter 5, Ribot and Larson warn that REDD+ may exacerbate forest people's political exclusion, although they recognize that it may open up new avenues for including forest people in democratic decision-making. MacKay, in Chapter 3, points out that REDD+ may remove some of indigenous peoples' legal gains over the past two decades, yet may also offer forest people new forums for the recognition of their particular identities, histories, and visions.

Forest rights activists have begun to respond to the challenge of defending forest people's existing rights and helping them to gain new rights under REDD+. A key demand has focused on institutionalizing forest people's participation in REDD+ decision-making on the basis of the United Nations Declaration on the Rights of Indigenous Peoples (UNDRIP). This demand made its way into the text passed at the Conference of Parties in Cancun at the end of 2010. The text also refers to the knowledge of indigenous peoples and the members of local communities. Other demands have not been as successful, such as calls for a fair distribution of REDD+ benefits, the recognition of forest people's tenure rights to forest, and the transfer of carbon rights to forest people.

In this concluding chapter, we discuss the implications of this book for efforts to promote forest people's rights in times of REDD+. We begin by synthesizing the main findings of the analyses presented in the preceding chapters with regard to REDD+, highlighting unity in diversity as a defining characteristic of the forest rights agenda. Building on this synthesis, we then identify four key challenges for forest rights activists. First, forest people and their supporters need to strengthen their global unity, as the increasing significance of global conventions, transnational regulations, and North–South connections require novel forms of mobilization and advocacy. Second, while enhancing global unity, activists need to sustain the diversity that has made the forest rights agenda so vibrant and relevant this far. They will only avoid reliance on silver bullets if they preserve the capacity of reflexive recognition pointed out in the introduction. Third, forest people and their supporters need to develop strategic representations in support of transnational mobilization and advocacy. Reference to "forest commons" may help them to highlight key features of the forest rights agenda. Fourth, forest rights activists need to employ the notion of forest commons to critically engage other rights-centered agendas that do not share the goal of social justice.

Unity in Diversity

The single most important message emanating from this book is that there is a vibrant and distinguishable forest rights agenda. Forest rights activists share a unifying vision centered on strong commitment to social justice, as highlighted in Chapter 1 of this volume, and apparent from virtually all contributions to this book. Social justice provides a unifying vision that helps forest rights activists to bring together highly diverse demands and make productive use of the conceptual tensions encountered in their work. The commitment to social justice finds concrete expression in three central demands generally shared by activists: the redistribution of forest tenure; forest people's participation in political decision-making; and the recognition of their particular identities, experiences, and aspirations.

Forest rights activists find rights a useful concept to promote forest people's empowerment. The reference to rights is malleable enough to capture demands focusing on property rights, participation in governance, and recognition of identities. Rights are also sufficiently plastic to allow articulations at local, national, and global levels, as we have argued in Chapter 1. At the local level, forest people's demands refer to concrete bundles of rights and duties regarding specific forest resources and functions. National rights claims take the form of legal relationships and procedures that apply uniformly across national territories. Global assertions find their expression in generalized notions of moral entitlement.

As a result, forest rights cannot be defined in a uniform and universal manner. The claims made by forest people at local, national, and global levels are highly diverse, as they refer to different kinds of rights, sorts of actors, and types of authority, and call for different political strategies. Rights, in other words, are not "out there" waiting to be recognized, nor can they simply be imposed from outside through legislation or other means. Rights claims are always specific to the political economic and ecological settings in which they are made, and their recognition as rights only gains concrete meaning within particular economic, political, cultural, and ecological conditions.

The contributions to Part II of this book demonstrate this malleability or plasticity of rights. They show how abstract calls for forest rights take highly varied manifestations in concrete settings, as people articulate their demands in diverse ways. Ribot and Larson, in Chapter 5, compare the limited administrative rights that forest people receive under regulatory reforms with the broader political rights required for their empowerment. In Chapter 6, Barnhart contrasts the tenure rights that forest people have received through legislative reform with people's broader demands for political and human rights. To Xuan Phuc (Chapter 7) distinguishes the legal transfer of tenure rights from broader efforts to enhance people's economic rights, strengthening their capacities to translate tenure rights into tangible benefits.

The chapters in Part III demonstrate the multitude of actors making claims on forests. Forest people include a variety of individual and collective actors, who are differentiated by wealth, power, gender, ethnicity, and other factors. Moeliono and Limberg (Chapter 8) illustrate how the diversity of local actors typically often defies simplistic notions of "local forest communities" or "indigenous peoples", each sort of actors asserting valid justifications for their own claims on different moral grounds. In Chapter 9, Edwards shows how social actors change over time and calls for attention to power differences among them. Middleton (Chapter 10) demonstrates how exclusive rights definition may inadvertently favor some claimants over others. In addition, MacKay (in Chapter 3) adds that rights are not available to individual people only, but may also refer to forest peoples' collective entitlements.

Part IV shows that, in addition, rights may refer to different types of authority (i.e. different institutions sanctioning claims on forests as rights and different procedures endorsing forest rights). In Chapter 11, Larson and Cronkleton suggest that there are typically various state and customary institutions available for representing the people making claims on the forest and holding forest rights. Some of the relevant institutions may also operate at transnational or global levels, as indicated by Campese and Borrini-Feyerabend (Chapter 4). All these institutions tend to apply different procedures for accepting claims on forests, endorsing them as rights, and enforcing recognized rights, as pointed out by Dorondel in Chapter 12.

Just as concrete claims refer to various kinds of rights, sorts of actors, and types of authority, they also tend to call for diverse kinds of strategies to promote their recognition as rights (Part V). In Chapter 13, Singh shows how forest activists benefit from mobilizations at various scales and through various forums. Ratner and Parnell (Chapter 14) demonstrate how strategic coalitions with national and international non-governmental organizations (NGOs) serve local actors to get forest rights recognized. In Chapter 15, Cronkleton and Taylor find that a focus on political action needs to be complemented with other mobilizing strategies to maintain political momentum.

Despite this diversity, forest people and their supporters have begun to organize at the global level. They use transnational law and call upon transnational courts to promote the recognition of indigenous peoples' rights to land, forest and territory (Chapter 3). They develop human rights-based approaches to inform the worldwide operations of international NGOs and development organizations (Chapter 4). They also consult with local and national forest rights activists all around the globe and support their efforts through global policy assessments (Chapter 2). Furthermore, they perform global-level analyses of forest tenure reforms and forest governance (Chapter 11; see also Ribot's broader work).

Achieving Global Unity

These ongoing efforts of global organizing show that forest rights activists have recognized the need to strengthen global unity. Nevertheless, REDD+ challenges them to develop novel forms of global mobilization and advocacy. Global organizing poses a formidable challenge to forest rights activists for the conceptual reasons pointed out above. In the absence of a uniform right to forest, there is no single operational demand behind which forest rights activists can rally. More importantly, any attempt to develop such an operational demand endangers an essential feature of the forest rights agenda, a feature that has given the agenda much vibrancy and strength over the past years: strong roots in local and national mobilizations. These conceptual and strategic orientations turn efforts to achieve global unity into a tremendous challenge.

Yet, global mobilization and advocacy will be vital for the continuing relevance of the forest rights agenda. In the most immediate sense, global advocacy is important to lobby for references to forest people's rights in global, transnational, and national REDD+ regulations. In addition, global mobilization is imperative as a way to match the global scale of economic forces unleashed by the inclusion of forests in climate change mitigation. Local efforts of rights recognition will benefit from outside support when they encounter project developers roaming tropical forests, seeking to capture profits and/or carbon, whether they are "carbon cowboys" or environmental NGOs. National advocacy on the behalf of forest people's rights will gain strength through suitable collaborations and exchanges with parallel efforts in other countries. Transnational advocacy and mobilization will also help to improve Northern consumers' awareness of the actual or potential effects of REDD+ actions on forest people, and to convince them and their governments to support REDD+ actions beneficial to forest people. Finally, advocacy at the global or transnational level will serve to enhance attention to forest people's experiences and perspectives in research on carbon forestry concentrated in an increasingly small number of institutions around the globe.

The emergent globalization of the forest rights agenda necessarily involves processes of unification around a limited number of organizations, associations, networks, and forums. The number of organizations active in global and transnational advocacy, such as the Forest Peoples Programme, the Rights and Resources Initiative, and Tebtebba, will be small. There will be only a few associations and networks, such as the International Alliance of Indigenous and Tribal Peoples of Tropical Forests and Global Alliance of Community Forestry, gaining global status. Global forums, such as the Forest Day organized in parallel to the annual Conference of the Parties to the United Nations Framework Convention on Climate Change (UNFCCC), may attract increasing numbers of participants from all over the world. Research and policy advice will become increasingly centered on global concerns, as already observable at the Center for International Forestry Research (CIFOR).

In addition, global mobilizations may require forest rights activists to reach out to other social movements. Recent experience with global negotiations over the role of forests in climate change mitigation indicates the limits of an agenda specialized on forest rights. Forest rights activists initially watched at the sidelines when climate change negotiators turned to forests and put REDD on the table. Such experiences indicate the political advantages of a broader rights agenda that develops strategic alliances with related movements. Forest rights activists will need to consider the value of potential alliances, particularly with the climate justice movement and sympathetic environmental groups.

Sustaining Diversity in Unity

REDD+ challenges forest rights activists not only to strengthen global unity, but also to sustain their appreciation of the diversity of rights claims to forests. Unity needs to come without uniformity, as forest rights activists mobilize globally while acknowledging their roots in local and national mobilizations. After all, local and national efforts have provided the forest rights agenda with strength and vibrancy, as argued in Chapter 1 and highlighted by the emphasis on local and national experiences in this book. Forest rights activists need to preserve their appreciation of diversity if they want to sustain political momentum into the future.

When engaging REDD+, forest right activists need to assert different kinds of rights claims at global, national, and local levels. The reason for this is that, in the absence of a universal right to forest, forest people's demands will always be specific to the level of engagement and particular political economic context. Moreover, global demands will have to be sufficiently broad to accommodate variation across countries and localities. National rights claims will have to be flexible enough to accommodate variation in the specific claims made by forest people in specific settings. In the end, it is only at the local level that forest people's demands take place in terms of concrete bundles of rights and duties regarding specific forest resources and functions. Only when global definitions, national assertions, and local claims match to a sufficient degree will shared and robust definitions of rights emerge on the ground.

In other words, forest rights activists need to avoid reliance on silver bullets in global mobilizations. They need to accept that there is no single operational right to forests, and acknowledge that any attempt to rally forest rights activists behind any such rights claim would threaten a critical feature of the forest rights agenda. This insight applies to efforts to define the kinds of rights to which forest people should be entitled. Narrow focus on, for example, carbon rights would not do justice to the diversity of rights claims made by forest people at various levels and in diverse settings. Similarly, forest rights activists need to remind themselves that the right to be consulted under free, prior, and informed consent (FPIC) refers to a particular form of collective participation in a particular set of decisions only. As flexible and encompassing as the right may potentially be, it may not match forest people's actual claims in particular settings.

The need to avoid uniform operationalizations of forest rights possesses direct relevance to attempts to define the actors entitled to forest rights narrowly and then quantify their number. It makes as little sense to quantify the total number of people directly dependent upon forests in the world – 1.6 billion by a World Bank count from 2004 – as to lump various kinds of actors into single, uniform categories, such as "local forest communities" or "indigenous peoples". Some forest people may assert indigenous identities for themselves and thereby want to be considered as indigenous peoples; yet, equating forest people with indigenous peoples creates problems, as we have discussed in the introduction to Part III. The same reservations apply to any narrow definitions of relevant authority, such as sole reliance on transnational human rights conventions, or of effective political strategies, such as legal empowerment and court actions.

Forest rights activists can avoid reliance on silver bullets and sustain their appreciation of diversity if they maintain the capacity for reflexive recognition highlighted in Chapter 1. This capacity to recognize internal differences and engage in deliberative processes allows them to make productive use of tensions between different kinds of claims and understandings of forest rights. The capacity also allows them to develop different definitions of forest rights at multiple levels and, if necessary, adjust them over time. It is the basis of context-specific engagement in which global and national rights assertions work to support local rights claims instead of superseding them. Sustaining the capacity may take particular effort and potentially weaken the global "punch" of the forest rights agenda. Yet, it will be important to maintain the political momentum gathered by forest rights activists.

Global activists will need to go out of their way to maintain deliberative processes across levels of mobilization (i.e. with national movements and local initiatives). Cross-level communication implies that deliberations at any particular level involve activists from the other levels. In this way, forest rights definitions consider the nature of claims and understandings of rights established at various levels. The definition of rights at higher levels gains meaning by including effective participation from lower levels, particularly local voices. Cross-level deliberation thereby offers space for activists to voice their demands at more than one level, providing additional opportunities for inclusive definitions of forest rights. Cross-level communication will obviously require significant efforts from activists at all levels, but particularly by global activists to involve local and national activists in global strategies.

Developing Strategic Representations

The third challenge faced by forest rights activists is to develop strategic representations of their agenda. Strategic representations will serve them to promote possibilities for forest people's empowerment in national and global discussions about REDD+. The representations need to be strategic in the sense that they serve activists to defend forest people's interests and to advance their claims. They do not claim to depict reality, nor do they seek to deflect attention away from the local specificity, historical contingency, and political nature of forest rights claims on the ground. Articulating them in global debates about REDD+ can produce strategic gains for forest people, as they open up new opportunities by legitimating alternative courses of action. They also serve to counter dominant discourses, such as carbon-centered representations of climate change.

Reference to "forest commons" may help forest rights activists to communicate their vision of social justice. At the most fundamental level, speaking about forest commons calls attention to the fact that forests tend to be "peopled", that forests are not *terra nullius* (no man's land), open access, or up for grab. More specifically, calls for the recognition of forest people's rights to forest commons allow activists to emphasize inherently collective features of forest governance and forest people's identities, in addition to individual elements. The notion of forest commons also highlights the local variability and specificity of rights next to the presence of larger collectivities, such as nations or global society, with a stake in forests. In addition, rights to forest commons call for attention to issues of distribution, participation, and recognition.

Forest rights activists can employ the reference to forest commons to highlight the diversity of relevant rights, actors, authorities, and strategies. Relevant rights comprehend ownership rights, use rights, consultation rights, cultural rights, procedural human rights, etc. Actors include indigenous peoples and local forest communities, as well as individual people differentiated by wealth, power, gender, ethnicity, social identities, place of residence, etc. Authority over forest commons is shared between the nation-state and a variety of other politico-legal institutions, including customary institutions at the local level. Strategies to achieve rights recognition speak to a variety of forums at multiple levels through diverse means.

The notion of forest commons emphasizes the significance of negotiation and need for political strategies to promote rights recognition: forest rights are not "out there" waiting to be recognized. Instead, claims on forests and their recognition as rights are negotiated through political struggles conditioned by particular histories, biophysical landscapes, and cultural meanings. By implication, calling for "rights to forest commons" is different from demands for collective ownership. The reference to "commons" does not translate into a simple endorsement of collective actors, such as local forest communities or indigenous peoples. Neither does it involve the wholesale endorsement of customary institutions irrespective of actual authority relations on the ground. Instead, the reference to forest commons opens up grounds for forest people's empowerment. The reference emphasizes the significance of collective decision-making, collective definitions of what claims are recognized as rights, and recognition of collective powers to enforce and adjust rights, as well as arbitrate in disputes over forest rights through context-specific means.

Moreover, the notion of forest commons resonates with the key findings of this book. It speaks to the significance of broader political and economic rights in settings characterized by entrenched political and economic inequality, as highlighted in the introduction to Part II. The concept of commons matches the suggestion in the introduction to Part III that inclusive rights definitions possess advantages in situations where differently positioned actors compete for recognition of their claims on the basis of different justifications. It also fits the observation in the introduction to Part IV that emphasis on nested forms of authority may be superior to a focus on single institutions or procedures.

Engaging Other Rights-Centered Agendas

The fourth challenge to forest rights activists is to critically engage with other rights-centered agendas that do not share the goal of social justice. Reference to "forest commons" may help them to distinguish their own from other rights agendas of direct or indirect relevance to REDD+, such as discourses centered on private ownership or compensation for forgone opportunities. These discourses are typically characterized by narrow and universal definitions of rights, whether those are tenure, human, or legal rights. They tend to employ universal operationalizations of rights, deflecting attention away from their local specificity, ignoring their political nature, and looking away from the broader processes constituting rights. This universal treatment makes other rights-centered discourses more amenable to the demands for global policy-makers than the rights agenda as they can be translated more easily into operational policy and project guidelines.

Emphasis on forest commons will help forest rights activists to distinguish their agenda from neoliberal discourses centered on private ownership. The clarification of forest tenure, codification of carbon rights, and extension of private ownership are critical preconditions for the establishment of market-based governance. REDD+ has already attracted significant interest by private investors and the financial sector, as illustrated by George Soros's visit to Cancun in late 2010 to express his support for a global REDD+ agreement. "Carbon cowboys" are traversing the tropical world in the search for lucrative deals along the forest carbon frontier. Their interest in the transfer of property rights, even to forest people, is radically different from forest rights activists' demands for redistribution. In the absence of more encompassing political-economic changes, experience from land reform demonstrates that resource rights are as quickly lost by local people as received by them. The reference to forest commons would thus serve to distinguish the individualizing tendencies of neoliberal property discourses from forest rights activists' recognition of individual and collective elements and their attention to broader economic and political empowerment.

Similarly, emphasis on forest commons challenges discourses about forests as public goods to be managed by central governments for the national good. The concept of public goods has long served centralized forest departments to assert control over forests, exclude local people, and extend state power into remote areas. It is now being revived in the form of a REDD+ architecture founded on national REDD+ programs. Forests are once again defined as part of the national heritage, this time not for the timber they produce, but the carbon stocks they contain. REDD+, therefore, may serve national governments once more to define forests as a national treasure and to strengthen their control over valuable forest resources. Governments may be ready to discuss the ways in which the benefits from REDD+ may be distributed among involved actors; but most talks of benefit distribution are radically different from forest rights activists' demand for the redistribution of forest tenure and fair distribution of benefits and costs. Reference to forest commons would highlight that forest people's demands cannot be reduced to shares in financial revenues or compensations for the financial costs of forgone alternative forest uses. Instead, forest people demand tenure rights to forests, and they call for an equitable distribution of all benefits derived from forests and costs arising from forest use.

Reference to forest commons also sets the forest rights agenda apart from participatory discourses blacking out the entrenched inequalities characterizing most forest areas in the developing world. Significant efforts are currently being made to develop operational procedures for project-level consultations with forest people, facilitated by transnational human rights conventions, particularly UNDRIP. The operationalization of the principle of FPIC and related efforts have the potential to go beyond previous attempts of making forest governance more participatory or involving various kinds of actors into multi-stakeholder negotiations. Nonetheless, a narrow focus on project-level interventions fails to recognize the significance of higher-level legal and regulatory frameworks, as well as the effects of power differences on forest people's ability to voice their demands and to assert accountability. Local-level consultations may easily ignore larger issues, such as the question of property rights, land classification, and local democracy. In addition, local-level consultations may be appropriated by powerful actors to manipulate forest people. In contrast, the notion of forest commons would highlight the problematic nature of statutory law and power relations in many contexts. It puts the spotlight on forest people's broader demands for the recognition of their particular histories, identities, and visions of desirable forest landscapes.

Finally, we want to highlight one more advantage of reference to "forest commons". Forest rights activists cannot only improve the outward representation of their agenda by embracing rights to forest commons, thereby opening up possibilities for forest people's empowerment in national and global discussions about REDD+. They can also draw on the notion to remind themselves of the need to sustain their own capacity for reflexive recognition. Forest rights activists can employ the reference to forest commons to highlight the significance of deliberative processes amongst themselves – and to acknowledge the role of engaged research in critical and constructive reflection.

Index